Vitória Tereza Negrão de Albuquerque
Adna Cristina Barbosa de Sousa
Kristerson Reinaldo de Luna Freire

AXENIC CULTIVATION OF PLEUROTUS OSTREATUS VAR. FLORIDA

Vitória Tereza Negrão de Albuquerque
Adna Cristina Barbosa de Sousa
Kristerson Reinaldo de Luna Freire

AXENIC CULTIVATION OF PLEUROTUS OSTREATUS VAR. FLORIDA

AXENIC CULTIVATION AND NUTRITIONAL EVALUATION OF PLEUROTUS OSTREATUS VAR. FLORIDA, PRODUCED ON LIGNOCELLULOSIC SUBSTRATES

ScienciaScripts

Imprint
Any brand names and product names mentioned in this book are subject to trademark, brand or patent protection and are trademarks or registered trademarks of their respective holders. The use of brand names, product names, common names, trade names, product descriptions etc. even without a particular marking in this work is in no way to be construed to mean that such names may be regarded as unrestricted in respect of trademark and brand protection legislation and could thus be used by anyone.

Cover image: www.ingimage.com

This book is a translation from the original published under ISBN 978-620-6-75952-2.

Publisher:
Sciencia Scripts
is a trademark of
Dodo Books Indian Ocean Ltd. and OmniScriptum S.R.L publishing group

120 High Road, East Finchley, London, N2 9ED, United Kingdom
Str. Armeneasca 28/1, office 1, Chisinau MD-2012, Republic of Moldova, Europe
Managing Directors: Ieva Konstantinova, Victoria Ursu
info@omniscriptum.com

Printed at: see last page
ISBN: 978-620-8-11248-6

Copyright © Vitória Tereza Negrão de Albuquerque, Adna Cristina Barbosa de Sousa, Kristerson Reinaldo de Luna Freire
Copyright © 2025 Dodo Books Indian Ocean Ltd. and OmniScriptum S.R.L publishing group

ACKNOWLEDGMENTS

I would like to thank my mother, Maria do Socorro Negrão Costa, for her unconditional support. I am very grateful to all the friends who have strengthened, added to and are like family: Gabriela, Maria, Emília, Charlie and Vic.

I would like to thank Prof.ª . Dr.ª . Adna Cristina Barbosa de Sousa for her guidance and example as a professional and person. To Prof. Dr. Kristerson Reinaldo de Luna Freire for his co-supervision and to the other professors at the Biotechnology Center.

To the members of the Evaluation Board, Professors Dr. Ian Porto Gurgel do Amaral (DBCM/CBIOTEC/UFPB) and Dr.ª. Andréa Farias de Almeida (DB/CBIOTEC/UFPB) for agreeing to participate in the evaluation of this work and for all their valuable suggestions and contributions.

To the Biotechnology Center, together with all its employees and technicians, for providing the physical structure necessary to carry out the work. As well as the Molecular Genetics and Plant Biotechnology Laboratory (LGMBiotec/UFPB); Food Technology Laboratory (LTA/UFPB) and Plant Tissue Analysis Laboratory of the Agricultural Sciences Center of *Campus* II/UFPB, which were essential for carrying out this work.

To CNPq (National Council for Scientific and Technological Development) for the student funding, without which it would not have been possible to carry out this research.

SUMMARY

Pleurotus spp. are easy to cultivate due to their lignocellulosic characteristics. This allows it to colonize various compounds abundant in the agro-industry. Based on this, the aim of this study was to evaluate the biological and production parameters and to characterize the physicochemical properties of the basidiomata of *P. ostreatus* var. *florida* grown in different formulations based on bean pods without the grain (VF), malt bagasse (BM), sawdust (SE) and algaroba leaves (FA). Initially, a test was carried out on the influence of light and temperature on the mycelial development of 3 base formulations called TLT1, TLT2 and TLT3 made up of Rice (98%) + Agricultural Gypsum (2%); Malt Bagasse (100%) and Malt Bagasse (96%) + Agricultural Gypsum (2%), respectively. This was followed by a qualitative pre-selection of 14 new formulations for *spawn* and basidiom production tests. To do this, the individual substrates were assessed for pH, reducing sugars, non-reducing sugars and total sugars, ash and soluble solids. The carbon (C), nitrogen (N) and C/N ratio of the selected formulations were also assessed: Green Bean Pod without Grain (VF); Green Bean Pod without Grain + Agricultural Gypsum (VF+GA); Malt Bagasse + Green Bean Pod without Grain (BM+VF); Malt Bagasse + Green Bean Pod without Grain + Agricultural Gypsum (BM+VF+GA); Malt Bagasse + Algaroba Leaves (BM+FA); Grainless Green Bean Pods + Algaroba Leaves (VF+FA) and Grainless Green Bean Pods + Algaroba Leaves + Agricultural Gypsum (VF+FA+GA). The evaluation of the influence of light on mycelial growth showed that in the dark (24h) and photoperiod (12h light/12h dark), the mushroom showed better vegetative development. Treatment TLT1 (3.5) showed the greatest growth, followed by TLT2 (2.96 cm) and TLT3 (2.7 cm). The ideal temperature for vegetative growth in the three treatments was between 20°C and 25°C. The physicochemical tests of the isolated substrates indicated that malt bagasse (BM) had 4.5% total sugars and bean pods without grains (VF) 3.62%. It was not possible to determine the total sugar content of algaroba leaves (AF) with the methodology used. The BM had 3.41% ash, VF 2.24% and FA 5.55%. Moisture was 8.09%, 6.9% and 22.10% for BM, VF and FA, respectively. Brix was 1° (BM and VF) and 0.1° (FA). Eight formulations (VF), (FA+GA), (BM+VF), (BM+FA), (BM+VF+GA), (BM+FA+GA); (VF+FA) and (VF+FA+GA) showed density, vigor and continuous growth, so they have potential for producing *spawn* of *P. ostreatus* var. *florida*. Seven treatments were selected for production and formed basidiomes characteristic of the species. Colonization time was 15.14 days and the total cycle was 22.14 days for all treatments. The highest yield percentage was 154.76% (VF) followed by 87.39% (VF+GA) and the lowest were 71.05% (BM+VF) and 53.27% (VF+FA +GA). The highest biological efficiency (EB%) was 38% (VF) and 37.71% (BM+VF), while the lowest was 16.38% (VF+FA+GA) and 27.58% (BM+FA). The basidiomes produced in the 7 treatments had a low lipid value of 2 to 3%, a high protein content of 29 to 34% per 100g of dry mass, ash content of 3 to 4%, carbohydrates of 50 to 59% and an average caloric value of 350 Kcal. The C/N ratio of the 7 formulations was 26:1 (VF), 17:1 (VF+GA and BM+VF), 13:1 (BM+VF+GA and BM+FA), 14:1 (VF+FA) and 15:1 (VF+FA+GA). The best formulations in terms of yield and protein content are VF; BM+FA; BM+FA and VF+FA. In this way, the formulations with the BM, VF and FA substrates were conducive to *spawn* production and the development of edible mushrooms (Shimeji-white) with a high nutritional value.

Keywords: edible mushrooms, agro-industrial waste, fungal protein.

ABSTRACT

Pleurotus spp. cultivation is facilitated due to its lignocellulosic characteristics, allowing for high colonization capacity in various abundant compounds in agroindustry. Based on this, the objective of the present study was to evaluate the biological and productive parameters and to characterize from a physicochemical standpoint the basidiomata of P. ostreatus var. florida cultivated in different formulations based on bean pod without grain (BP), malt bagasse (MB), sawdust (SD), and carob leaves (CL). Initially, a test was conducted on the influence of light and temperature on the mycelial development of three base formulations named TLT1, TLT2, and TLT3 composed of Rice (98%) + Agricultural Gypsum (2%); Malt Bagasse (100%); and Malt Bagasse (96%) + Agricultural Gypsum (2%), respectively. Subsequently, a qualitative pre-selection of 14 new formulations was carried out for spawn and basidioma production tests. For this, individual substrates were evaluated for pH, reducing sugars, non-reducing sugars, total sugars, ash, and soluble solids. The carbon (C) and nitrogen (N) content and C/N ratio were also evaluated for the selected formulations: Bean Pod without Grains (BP); Bean Pod without Grains + Agricultural Gypsum (BP+AG); Malt Bagasse + Bean Pod without Grains (MB+BP); Malt Bagasse + Bean Pod without Grain + Agricultural Gypsum (MB+BP+AG); Malt Bagasse + Carob Leaves (MB+CL); Bean Pod without Grains + Carob Leaves (BP+CL); and Bean Pod without Grains + Carob Leaves + Agricultural Gypsum (BP+CL+AG). The evaluation of light influence on mycelial growth showed that in darkness (24h) and photoperiod (12h light/12h dark), the mushroom exhibited better vegetative development. Treatment TLT1 (3.5) showed the highest growth, followed by TLT2 (2.96 cm) and TLT3 (2.7 cm). The ideal temperature for vegetative growth in the three treatments was between 20°C and 25°C. Physicochemical tests of isolated substrates indicated that malt bagasse (MB) had 4.5% total sugars and bean pod without grains (BP) had 3.62%. It was not possible to determine the total sugar content in carob leaves (CL) with the methodology used. MB had 3.41% ash, BP 2.24%, and CL 5.55%. Moisture content was 8.09%, 6.9%, and 22.10% for MB, BP, and CL, respectively. Brix was 1° (MB and BP) and 0.1° (CL). Eight formulations (BP), (CL+AG), (MB+BP), (MB+CL), (MB+BP+AG), (MB+CL+AG), (BP+CL), and (BP+CL+AG) showed density, vigor, and continuous growth, therefore, they have the potential for P. ostreatus var. florida spawn production. Seven treatments were selected for production and formed characteristic basidiomata of the species. Colonization time was 15.14 days, and total cycle was 22.14 days in all treatments. The highest yield percentage was 154.76% (BP) followed by 87.39% (BP+AG), and the lowest were 71.05% (MB+BP) and 53.27% (BP+CL+AG). The highest biological efficiency (BE%) was 38% (BP) and 37.71% (MB+BP), and the lowest values were 16.38% (BP+CL+AG) and 27.58% (MB+CL). Basidiomata produced in the seven treatments showed low lipid content of 2 to 3%, high protein content of 29 to 34% per 100g dry mass, ash content of 3 to 4%, carbohydrate content of 50 to 59%, and average caloric value of 350 Kcal. The C/N ratio of the seven formulations was 26:1 (BP), 17:1 (BP+AG and MB+BP), 13:1 (MB+BP+AG and MB+CL), 14:1 (BP+CL), and 15:1 (BP+CL+AG). The best formulations in terms of yield and protein content are BP; MB+CL; MB+CL and BP+CL. Thus, formulations with MB, BP, and CL substrates provided the production of spawn and development of the edible mushroom (White Shimeji) with high nutritional value.

Keywords: Edible mushrooms, agro-industrial waste, fungal protein.

SUMMARY IO

1 INTRODUCTION

Pleurotus ostreatus var. *florida* also known as Shimeji-white or oyster mushroom is the third most produced mushroom in the world, due to its suitability for subtropical and tropical climates and its ability to use a variety of lignocellulosic residues as a nutritional source through the action of enzymes such as lacase and peroxidase to promote its vegetative growth and fruiting (Thakur, 2020).

Currently, there is a high demand for more sustainable food products with a strong nutritional appeal, especially when it comes to plant-based protein. And in this scenario, mushrooms are beginning to be valued in the market due to their cholesterol-free nutritional attributes, high protein content and important essential and non-essential amino acids (Lessa *et al.*, 2022). In addition, there is growing scientific interest both in the food industry for application in functional foods and in the pharmaceutical sector in investigating secondary metabolites from agaricomycetes (Basidiomycota, Fungi) and their components with antioxidant, antimicrobial, antiviral, antitumor and immunomodulatory properties. Functions conferred by a variety of molecules made up of polysaccharides, terpenes, statins, peptides, phenolic compounds, carotenoids and other bioactive compounds (Doroski *et al.*, 2022; Inci *et al.*, 2023).

In this context, mushroom production has a high potential due to the axenic cultivation technique, which makes it possible to produce more protein per cultivated area. From axenic cultivation, the most varied lignocellulosic substrates can be used as a source of energy, which allows the use of materials discarded in the agro-industry, with the aim of making mushroom cultivation more economical.

A large amount of agricultural biomass is discarded every year, including residues from corn, wheat, rice, sorghum and banana leaves. These materials provide significant levels of lignin, cellulose and hemicellulose, which are useful components for various bioprocesses. In addition, these residues serve as a source of carbon, nitrogen and essential minerals, which are fundamental for fungal growth. The variety of nutrients plays a crucial role in mushroom production, where the nutritional source is key to guaranteeing the quality of the crop, both in terms of yield and nutritional properties (Babu *et al.,* 2022).

One of the promising strategies to valorize these materials is the recycling of non-consumable biomass. And in this context, axenic cultivation technology has emerged as a possibility, allowing the cultivation and production of various mushroom species (Grimm *et al.,* 2021).

Pine sawdust (*Pinus elliottii*) is often used as a substrate in mushroom cultivation, especially for species such as *Lentinula edodes* and *Pleurotus* spp. It is a popular choice for mushroom cultivation due to its availability, as it is a by-product of the timber industry and is widely available in many regions in an affordable form. In addition, it has a porous structure, water-holding capacity and is an easy-to-handle compost (Tavarwisa *et al.*, 2021). Malt pomace (*Hordeum vulgare* L.) accounts for 85% of the waste generated in the brewing industry, obtained after the mashing stage in which the malt is ground and cooked. Brazil is the third largest producer of beans (*Phaseolus vulgaris* L.) and the Northeast produces 788,000 tons a year. The bean pod is a residue obtained after harvesting and threshing the beans. The algarobeira (*Prosopis juliflora* (Sw) DC) is a tree species that was introduced to the sertão of Pernambuco due to its adaptation to high temperatures and has multiple applications ranging from wood for firewood and fencing, to uses as a food source for herds during the dry season, a resource for beekeeping and other applications, being essential for the subsistence of local communities (Santos *et al.*, 2020). Due to its high leaf biomass, it is a valid research alternative for growing edible mushrooms.

Parameters such as humidity, temperature, aeration and photoperiod, type of substrate, pH, C/N ratio and bioavailable nutrients such as sugars and minerals according to the type of lignocellulosic material can influence mushroom cultivation. In addition to the levels of cellulose, hemicellulose and lignin, which can vary depending on the type of plant material, and the quality of the spawn used for primary colonization in the first stage of fungiculture development (Muswati *et al.*, 2021). Therefore, in order to make better use of waste in mushroom production, it is necessary to investigate the influence of different parameters and locally available materials for the production and generation of products with high economic value, in accordance with the reality of the region, so that local waste can be satisfactorily transformed into value-added biomass using fungiculture as an alternative to diversify regional agriculture. Based on this, the aim of this study was to evaluate the biological and production parameters and to characterize the basidiomata of *P. ostreatus* var. *florida* grown in different formulations based on bean pods without the grain, malt bagasse, sawdust and algaroba leaves from a physical and chemical point of view.

2 THEORETICAL BACKGROUND

1.1. Biological aspects of *Pleurotus ostreatus*

The *Pleurotus ostreatus* (Jacq.) mushroom (Figure 01) is an Agaricomycete known as the white rot fungus because it occurs spontaneously on decaying tree trunks. It plays an important ecological role in breaking down complex organic materials into forms that are more assimilable by other microorganisms for the cyclical permanence of the elements in forest ecosystems. In its wild macroscopic form, it is commonly found on the ground in forests, gardens, backyards and around rice fields, in dead wood trunks and organic waste (Daud *et al.,* 2021).

Figure 01. *Pleurotus ostreatus* on a tree trunk.

Source: http://raizesefolhas.info

Morphologically, it can be identified by its smooth, flat, bivalve-shaped hat, irregularly shaped like a fan or oyster. The hats have a convex shape when young, flattening out and increasing in size as the basidium ages. The edges of the hat are also irregular and change over time. The lamellae that make up the hemisphere occupy the space between the edges of the hat and the stipe (Figure 02). The color can vary from white to gray or brown, depending on the variety and the stage of development of the mushroom. They generally grow on a platform of overlapping clusters and may not develop a stem when growing laterally. In addition, *P. ostreatus* var. *florida* has a delicate and mild flavor, sometimes described as similar to shellfish and its texture is firm and fleshy, and it can be eaten in a variety of ways (Alexopoulos *et al.,* 1996).

Figure 02. Structure of *Pleurotus ostreatus* var. *florida.*

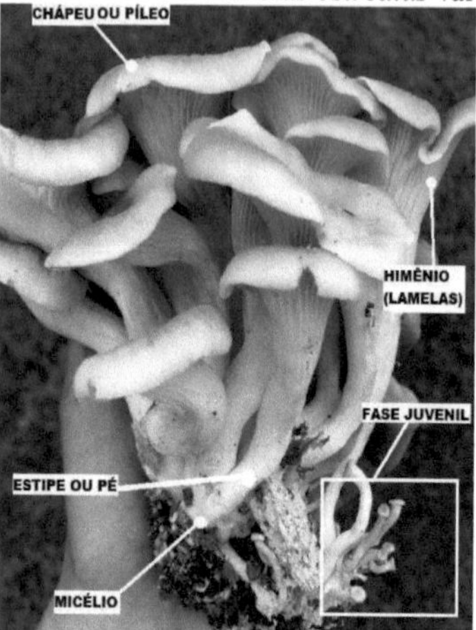

Source: http://cascadiamushrooms.com/education/

Cultivation of the species began to spread in Germany during the First World War as a solution to guarantee the subsistence of the population affected by food shortages. Its cultivation has been widespread because it has remarkable nutritional benefits and is easy to grow compared to other varieties of fungi, as it has a fast growth time, resistance to pests and diseases, the ability to thrive on a variety of cellulose residues and a set of enzymes for biological degradation that dismantle the structure of the plant cell wall, extracting the carbohydrates, lipids, proteins and minerals needed to sustain both its primary and secondary metabolic processes (Daud *et al.,* 2021). In this way, they use the structures of materials such as lignin, cellulose and hemicellulose commonly found in biodiversity and agricultural systems as their main carbon sources. They use the sugars present in these structures and, in the case of *P. ostreatus,* mainly consume glucose and fructose as their source of nutrients (Sun *et al.,* 2023).

The degradation of lignocellulose requires the synergistic action of multiple carbohydrate-active enzymes involved in breaking glycosidic bonds. Biomass degradation then occurs through the cooperative activities of hydrolytic and oxidative enzymes. The hydrolytic system is responsible for the degradation of cellulose and hemicellulose, while the oxidative system is known to participate in the degradation of lignin (Figure 03). It is one of the main groups of lignocellulolytic organisms, as they produce degrading enzymes such as cellulase, ligninase and hemicellulase. Lignin is bound directly (covalently) or indirectly (ester or ether) to cellulose and hemicellulose. These fungi produce enzymes such as laccase, lignin peroxidase (LiP) and manganese peroxides (MnP) that can break down even the toughest lignin compounds (Kumla *et al.,* 2012).

Figure 03. Enzymes produced by *Pleurotus ostreatus* .

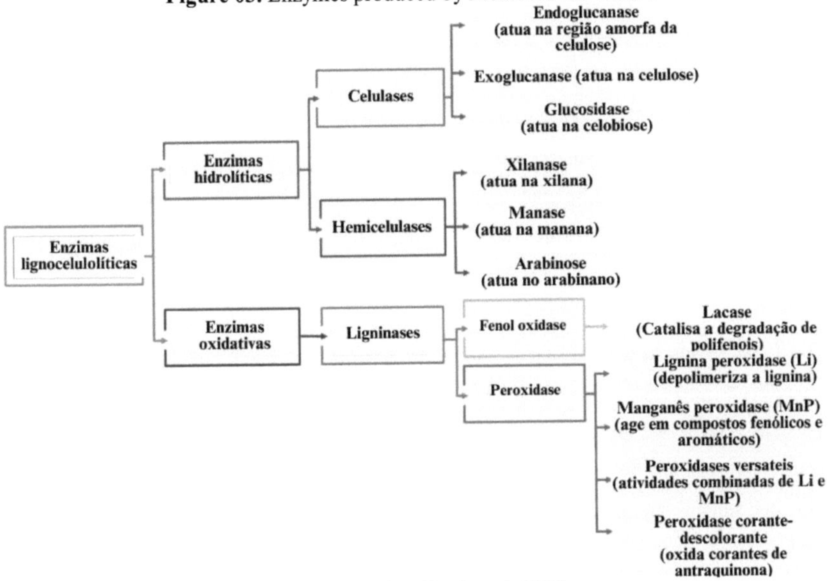

Enzimas lignocelulolíticas
Enzimas hidrolíticas
Celulases
Endoglucanase (atua na região amorfa da celulose)
Exoglucanase (atua na celulose)
Glucosidase (atua na celobiose)
Hemicelulases
Xilanase (atua na xilana)
Manase (atua na manana)
Arabinose (atua no arabinano)
Enzimas oxidativas
Ligninases
Fenol oxidase
Lacase (Catalisa a degradação de polifenois)
Peroxidase
Lignina peroxidase (Li) (depolimeriza a lignina)
Manganês peroxidase (MnP) (age em compostos fenólicos e aromáticos)
Peroxidases versateis (atividades combinadas de Li e MnP)
Peroxidase corante-descolorante (oxida corantes de antraquinona)

Source: Adapted from Kumla et al. (2020).

P. ostreatus has the highest activities of laccase and manganese peroxidase (responsible for lignin degradation), endoglucanase (for cellulose hydrolysis) and xylanase (hemicellulose hydrolysis). The species is able to reduce lignin more effectively than other similar fungi (Kuijk *et al.,* 2015).

1.2. Growing mushrooms

Mushroom cultivation begins on an organic base in a humid solid state, characterizing it as a low-cost Solid State Fermentation (SSF) cultivation method that is closer to the fungi's natural environmental conditions. In this process, the mycelium develops in a solid medium forming a plate of cell aggregates that covers the gas exchange surface, eliminating the need for mechanical agitation (An *et al.,* 2021). This method of cultivation on solid substrate has many advantages, such as the ability to reuse waste and convert it into high value-added bioproducts. This approach stands out for its sustainability and efficiency in mushroom production, taking advantage of materials that would otherwise be discarded. This ability to reuse the substrate adds additional value to mushroom production, as it contributes to the sustainable management of agricultural resources. The substrate used, after fulfilling its role in mushroom cultivation, can be used for animal feed, providing additional nutrients, or used as an organic fertilizer, promoting soil fertility. This practice benefits not only the mushroom farm itself, but can also be extended to third parties, expanding the positive impacts of this agricultural activity (Martin *et al.,* 2022).

Growing mushrooms using FES involves using a solid substrate as a medium for mycelial growth. This substrate can be made up of a variety of organic materials, such as agricultural waste, straw, sawdust, sugar cane bagasse, among others. The process generally follows these steps:

2.2.1 Preparation and handling of inoculum *spawn*: also known as "mushroom spawn", this refers to material colonized by fungal mycelium, specifically prepared for use as an inoculant. The *spawn* production process is usually carried out under sterile conditions to avoid contamination by unwanted microorganisms and guarantee the genetic purity of the cultivated fungus. The substrate must be composed of sterile material, such as cereal grains, sawdust, straw, among others, which are colonized by the mycelium of the selected species (Fornito *et al.,* 2020).

Currently, there are predominantly four categories of inoculum for mushroom cultivation: sawdust spawn, grain spawn, liquid spawn and stick spawn (Zhang *et al.*, 2019). When added to the main substrate, usually legume seeds, the mycelium quickly colonizes this substrate, which can be stored and transported to transfer the fungus to other substrates transforming them into a medium conducive to mushroom development. The quality of the *spawn* is fundamental to the success of mushroom cultivation, as a healthy and vigorous *spawn* will ensure rapid colonization of the substrate and efficient mushroom production. As the *spawn* is used to start the mushroom cultivation process and requires a controlled environment, mushroom spawn distributors play an essential role in delivering high-quality materials that improve productivity. For example, the demand for *spawn* in India is estimated at approximately 8,000 to 10,000 tons per year (Thakur *et al.,* 2020).

2.2.2 Preparation of the substrate: The substrate is prepared through a wetting process with humidity ranging from 65 to 75%. Generally, the substrates are moistened for 24 hours with distilled water and the excess water is removed, and sterilized when necessary, to eliminate any unwanted microorganisms that may compete with the fungi during cultivation (Fornito *et al.,* 2020). Depending on the substrate, it may need treatment such as pre-cooking or grinding because the smaller the particle size, the easier it is for the mycelium to penetrate the substrate block. This substrate can be made up of a variety of organic materials. Such as various industrial, agricultural and wood waste such as corn, wheat and bean straw, sawdust, sugar cane bagasse, among others (Muswat *et al.,* 2021).

2.2.3 Inoculation: After the substrate has been properly prepared, it is inoculated with selected fungal spores belonging to the desired species. This can be done through *spawn*, spores or by introducing a disk or small block of pure mycelium culture (Grimm *et al.,* 2021).

2.2.4 Incubation: After inoculation, the containers containing the inoculated substrate are incubated in the dark under suitable temperature and humidity conditions. During this period, the mycelium develops in the substrate. During its growth, the mycelium spreads throughout the substrate, secreting enzymes that degrade the organic matter present, transforming it into nutrients that can be absorbed by the fungi (Elkhateeb *et al.*, 2022).

2.2.5 Formation of primordia: When environmental conditions are suitable, such as humidity and temperature, the mycelium begins to form small structures called primordia, which will eventually develop into basidiomata, which are the mushrooms themselves.

2.2.6 Harvesting: The mushrooms are harvested when they have reached the desired size and maturity. Depending on the species grown, it can take several days or weeks from inoculation to harvest.

2.2.7 Post-harvest: After harvesting, mushrooms can be processed for immediate *fresh* consumption or stored dehydrated or preserved for future use. The remaining substrate can be used as fertilizer or animal feed, completing the sustainable cycle of mushroom cultivation (Zakil *et al.,* 2020).

1.3.Factors influencing productivity

2.3.1 Carbon/Nitrogen ratio: this is an important measure in mushroom cultivation and refers to the ratio between the amount of carbon and nitrogen present in the substrate used for fungal growth. It directly affects the availability of nutrients for the mycelium during the process of colonization and mushroom production (Elbagory et al., 2022).

An adequate C/N ratio in the substrate is essential to ensure healthy and robust mushroom growth. Generally, a C/N ratio of between 20:1 and 30:1 is considered ideal for many commercially cultivated mushroom species. Although this value can vary depending on the mushroom species and the type of substrate used. For *P. ostreatus*, better yields have been reported at ratios of 7:1 to 40:1 (Elbagory et al., 2022, Feng *et al.*, 2023).

When the C/N ratio is too high (more carbon than nitrogen), the substrate can be poor in nitrogen, which can limit mycelium growth and reduce mushroom production. On the other hand, if the C/N ratio is too low (more nitrogen than carbon), there may be an excess of nitrogen in the substrate, which can also be detrimental to mushroom growth and favor the growth of unwanted competing microorganisms (Desisa et al., 2023; (Bellettini *et al.*, 2019). Therefore, controlling the C/N ratio in the substrate is an important consideration in mushroom cultivation, and can be adjusted by properly choosing the materials used in the composition of the substrate and by adding nitrogen sources such as soybean meal, corn meal or manure if necessary (Melanouri *et al.*, 2022; Colla et al., 2023).

2.3.2 Moisture: this is the key factor in mushroom production and significantly affects fruiting/production. It is essential for the growth of the network of filaments that make up the structure of the fungi and directly affects their expansion and development in the substrate. The ideal humidity for growing *Pleurotus* spp is between 65 and 75%. Lack of humidity can lead to the formation of shriveled mushrooms, with irregular growth or deformations, while excess humidity can increase the risk of problems such as rotting and bacterial or fungal contamination (Navarro *et al.*, 2020).

2.3.3 pH: pH can affect the solubility of nutrients in the substrate, especially minerals which can be essential for the satisfactory growth of mushrooms, and can also affect the regulation of enzymes involved in fungal metabolism involved in the degradation of raw materials and assimilation of nutrients. Recent studies show that for the cultivation of *Pleurotus* spp. the ideal pH can vary between 4 and 6, slightly acidic (Vilas *et al.*, 2020). In industry, pH control is crucial to optimize production and minimize microbial competition (Nwoko *et al.,* 2021).

2.3.4 Temperature: higher temperatures generally accelerate the growth of the mycelium, but if they are too high, they can be harmful. On the other hand, lower temperatures can slow mycelial growth or inhibit it. It can affect colony morphology and reproduction (Zheng *et al.,* 2021). It is also related to the enzymatic activities of lacases, xylanases, amylases and proteases, which influence the advance of the mycelium on the substrate and fruiting (Zhuo *et al.,* 2018). This is directly linked to the quality of the mycelium and, consequently, the *spawn*. Among cultivated mushrooms, *Pleurotus* spp. is well adapted to the temperatures of tropical and subtropical climates, ranging from 25° to 30°C (Sen et al., 2020).

2.3.5 Light: light signaling regulates metabolic pathways in fungi and can trigger a molecular pathway, generating an oxidative stress response, which leads to suppressed mycelial growth in some species (Zheng *et al.,* 2021). There are species that grow in the dark and others in partial light. It seems likely that all mushrooms that require light use a common regulatory pathway for basidiom development. *P ostreatus* requires light for the formation of primordia and its reproductive structures (Belletini *et al.,* 2019). The direction and intensity of light can affect the growth and orientation of the stipe and pileus, i.e. they exhibit phototropism. In *indoor* cultivation systems, selected artificial light such as UV-B, red and blue light are used to enhance the quality of basidiomata. Light can also affect the production of bioactive compounds in mushrooms, such as vitamin D in mushrooms exposed to UV light. These compounds can have beneficial effects on health and can be influenced by exposure to light during cultivation. Mushrooms do not depend on lighting for their primary growth (myceliation), however certain degrees of lighting can affect mycelial vigor in artisanal cultivations (Belletini *et al.,* 2019).

1.4 Lignocellulosic waste

In the agro-industry, a variety of waste can be generated during the processing of agricultural products. Some of the main wastes discarded in the agro-industry include fruit and vegetable peels that are often discarded during food processing, such as apples, oranges, bananas, potatoes, among others. Vegetable leaves and stalks, such as cabbage, broccoli and lettuce, are often not used in the production process and are therefore discarded as waste (Ansiliero *et al.,* 2020). Sugarcane bagasse from the sugar and ethanol industry; straw and crop residues from crops such as rice, corn and wheat, straw and other crop residues are often left in fields or discarded as waste; bagasse and grain husks from grain milling industries such as wheat, corn and rice and from the brewing industry malt bagasse and other waste (Ricardino *et al.,* 2020).

These materials are characterized by a diversity of biocomposites, but the majority of their structural components are essential polymers such as cellulose, lignin and hemicellulose. The composition of these compounds in agro-industrial waste depends on the species, tissue and maturity of the plant. However, there is an average pattern to the balance of these compounds found in agro-industrial waste (Figure 04).

Figure 04. Balance of the average composition of common agro-industrial waste .

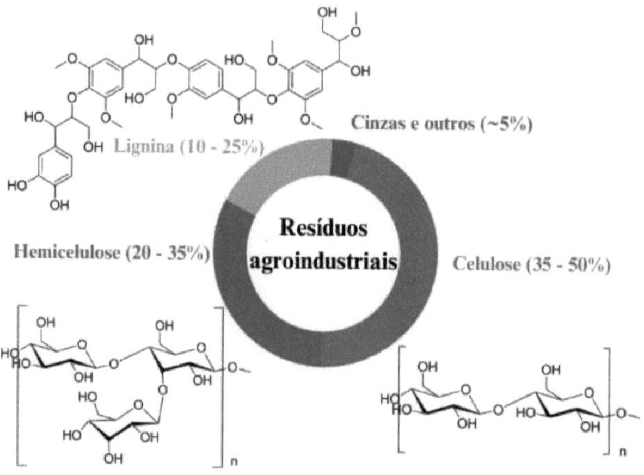

Source: (Adapted from Kumla *et al.*, 2020).

Cellulose is a homopolymer made up of repeated glucose units linked by β-1,4 bonds that have a β-anhydroglucose unit, with three hydroxyl groups (OH): a primary hydroxyl group at the C6 position and two secondary hydroxyl groups at the C2 and C3 positions. These hydroxyl groups are important because they can form intra- and intermolecular hydrogen bonds. Hydrogen bonds are intermolecular interactions between hydrogen attached to an electronegative atom, such as oxygen or nitrogen, and another electronegative atom. In cellulose, hydrogen bonds between the hydroxyl groups of adjacent glucose molecules are responsible for conferring stability to the cellulose structure and for its physical properties, such as strength and insolubility in water (Jedvert *et al.,* 2017). Saprophytic mushrooms have the ability to secrete enzymes, such as cellulases, which can break down the glucose bonds present in cellulose, releasing it for use as an energy source (Thakur, 2020).

Hemicellulose is a heteropolymer consisting of a polysaccharide structure. Its structure can vary considerably based on the sugar units present, the length of the chain and the branching of the molecules. Common sugars found in hemicellulose include pentoses such as xylose and arabinose, hexoses such as mannose, glucose and galactose, as well as hexenuronic acids such as 4-O-methyl-D-glucuronic acid, galacturonic acid and glucuronic acid. Small amounts of other sugars may also be present, as well as acetyl groups. These sugars can be grouped into different types of hemicellulose polysaccharides, including galactans, mannans, xylans, xyloglucans and β-1,3/1,4-glucans (Kumla *et al.,* 2020). Enzymes, including hemicellulases, xylanases, mannanases, arabinofuranosidases, among others, have the function of breaking the bonds between the sugars present in hemicellulose, thus releasing the monosaccharides to be consumed (Elkhateeb *et al.,* 2022; Colla *et al.,* 2022).

Lignin is a complex, rigid, amorphous and hydrophobic polymer. It is formed by a three-dimensional network of phenylpropanoid units, mainly monolignols, which are linked by different types of bonds, including β-aryl ether bonds, carbon-carbon (C-C) bonds and carbon-ether (C-O) bonds (Rico-Gárcia *et al.,* 2020). The structure of lignin is highly branched and has a variety of functional groups, including aliphatic, phenolic hydroxyl, carboxylic, carbonyl and methoxyl groups. These functional groups give lignin a diversity of chemical and physical properties, making it an extremely complex molecule (Kumla *et al.,* 2020). White rot fungi and brown rot fungi are efficient at degrading lignin and are often the main culprits in decomposing woody materials in nature. These organisms secrete oxidative enzymes such as laccases, peroxidases and others. They have the ability to break down lignin bonds into smaller, more assimilable molecules (Hongzhang, 2016).

This waste represents an opportunity to implement sustainable waste management practices, such as composting, biogas production, animal feed production, organic fertilizer production and other waste recovery processes. Especially for the production of other foods, such as edible and medicinal mushrooms (Zakil *et al.,* 2020). The efficient use of this waste can not only reduce the environmental impacts associated with its improper disposal, such as the emission of gases (methane and carbon dioxide) emitted from the anaerobic decomposition of waste in landfills; contamination of water and soil through eutrophication. But also in creating opportunities for income generation and promoting the circular economy in the agro-industry (Garlapati *et al.,* 2020).

To mitigate these environmental impacts, it is important to implement organic solid waste management practices that prioritize reduction at source, reuse, recycling and proper waste treatment (Garlapati *et al.,* 2020).

Mushrooms are efficient at recycling organic waste, and have the ability to transform these materials that could become pollutants into valuable sources of nutrition and substrates for cultivation. This characteristic makes mushrooms an attractive option not only for the production of healthy food, but also for sustainable waste management (Zakil *et al.,* 2020).

In this scenario, among the diversity of agro-industrial waste and lignocellulosic materials available in the northeast region, there is easy access to a diversity of sawdust from the region's timber sector; there are green bean pods, an agricultural crop present in the semi-arid region of Paraiba that generates easily accessible and low-cost biomass; the leaves of the algarobeira, a vegetation introduced to the hinterland of Pernambuco due to its adaptation to high temperatures, and malt bagasse from local breweries (Coelho; Ximenes, 2020; Santos *et al.*, 2020; Massardi *et al.*, 2020).

2.4.2 *Pinus* sawdust*:* the *Pinus* genus plays a significant role as a supplier of wood to various industries in Brazil, especially in the areas of pulp, lamination and sawmills. These trees are cultivated and exploited throughout the country, although they are more common in the South. However, like other wood species, *Pinus* is not fully utilized as a raw material in industrial processes. In Brazil, the wood industry often uses materials inefficiently, resulting in 15% to 50% of the raw material being lost during wood processing. This generates a significant amount of waste that could be put to better use (Lima *et al.,* 2019; Rugolo *et al.,* 2020). *Pinus elliottii* sawdust (Figure 05) contains components such as lignin, cellulose and hemicellulose, which are important for growing mushrooms and for other applications in the biomass industry. And, like other sawdust, *P. elliottii* sawdust has the ability to retain water, which is important for its use in various applications, including as a substrate for growing mushrooms (Mendes et al., 2021).

Figure 05. *Pinus Elliottii* wood sawdust.

Source: Battistella.com

2.4.2 Malt bagasse: in Brazil, the correct disposal of malt bagasse waste generated in the brewing industry is regulated by the National Solid Waste Policy (Law No. 12.305/2010) and other related environmental regulations. According to Brazilian legislation, companies are responsible for the proper management of all the waste they generate, including malt cake. It is important that brewing companies comply with the guidelines established by Brazilian environmental legislation and adopt waste management practices that promote sustainability and environmental responsibility. This includes implementing waste management programs, looking for recycling and reuse solutions, and minimizing waste generation whenever possible (Tombini *et al.,* 2024). Malt bagasse (Figure 06), which represents a large part of the waste generated by the brewing industry, is obtained after the mashing process, in which the malt or barley is ground and boiled. In Brazil, the average production of this waste is 2.82 million tons per year. In the beer production process, barley malt plays a crucial role, giving the drink its distinctive aroma and flavor. After fermentation, the malted barley grains are transformed into what is known as malt cake, making up around 85% of the total waste generated. This waste, basically made up of the malted barley husk, represents the main by-product of the brewing industry, available in large volumes throughout the year and at a relatively low cost. It is estimated that the production of every 1000 liters of beer results in 350 kg of wet waste (Tombini *et al.,* 2024). Although most of it is currently destined for animal feed, this practice has presented significant challenges, including storage problems and the risk of contamination by microorganisms that can cause disease in cattle, as noted by Massardi *et al.* (2020).

Figure 06. Malted barley grains *Hordeum vulgare L.*

Source: http://pertershiil.de

Barley is made up of the barley embryo (Figure 07. A), which is considered the fertile part of the seed and also contains starch, proteins and lipids. These are used during the development of the embryo after fertilization and as an initial source of nutrients when the seeds begin to germinate. By the starch granules incorporated into the protein matrix (Figure 07. B), made up of starch, amylose and amylopectin. Amylose is a linear polymer made up of glucose molecules linked by a-(1-4) glycosidic bonds. The ratio of amylopectin to amylose is around 3:1. And finally, the aleurone layer (Figure 07. C), which contains proteins, starch and lipids. Aleurone cell walls contain B-glucan, arabinoxylan, phenolic acids and storage proteins (Fox, 2010).

Figure 07. Structure of the barley/malt grain.

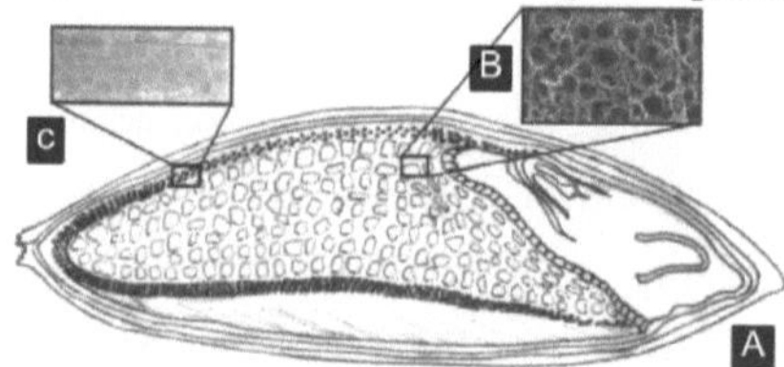

A: Embryo. B. Starch granules incorporated into protein matrix. C: Aleurone layer. Source: Fox (2010).

As stipulated by Brazilian regulations (Brasil, 2019a), the definition of beer encompasses the fermentation process conducted by yeasts of the genus *Saccharomyces*, using malted barley wort or malt extract, subject to a boiling phase with the addition of hops or hop extract. In addition, partial substitution of barley malt or malt extract with brewing adjuncts is permitted. In this context, beer is characterized as a carbonated drink, with an alcohol content of between 3 and 12% (v/v), made from a combination of barley malt, hops, yeast and water, which may include other elements such as rice, corn and wheat.

The composition of malt cake can vary depending on the commercial variety of malt and the type of brewing process. However, in general, it contains a number of important components, as observed by Qin *et al.* (2018) (Table 01).

Table 01. Composition of malt bagasse.

Component	Content found
Cellulose	16,8% - 20,6%
Hemicelluloses	18,4% - 28,4%
Lignin	9,9% - 27,8%
Proteins	15,3% - 26,6%
Carbohydrates	74,27%
Ashes	2,7% - 4,6%

Source: Qin et al. (2018) and Tombini et al. (2024).

These data highlight the richness in nutrients and compounds of malt pomace, making it a potential source for growing edible mushrooms (Massardi *et al.*, 2020).

2.4.3 Green bean pods: the Northeast is the largest producer of green beans in Brazil, producing approximately 788,000 tons per year (Coelho; Ximenes, 2020). In the semi-arid region of Paraíba, bean cultivation plays a significant role in generating biomass, especially in the form of bean pods (*Phaseolus vulgaris* L.) (Figure 08). After harvesting and threshing the beans, the straw and pods are obtained as residue, representing an easily accessible and low-cost source of biomass, as observed by Lima *et al.* (2019).

Figure 08. Bean pods and grains (*Phaseolus Vulgaris* L.).

Source: http://agroresidues.hg.

The green bean pod is characterized by an average moisture content of 87.31%, as well as containing a high protein content and significant amounts of potassium, phosphorus and calcium (Kalili *et al.,* 2022). When considered on a dry basis, its composition includes several important components as illustrated in Table 02. The nutritional and structural composition makes this waste suitable for application as a nutritional source in the cultivation of fungi of commercial interest, with potential for the production of edible mushrooms.

Table 02. Composition of green bean pods (*Phaseolus vulgaris L.).*

Component	Content found
Cellulose	33%
Hemicelluloses	46%
Lignin	17%
Proteins	29%

Source: Kalili *et al.* (2022).

The use of bean pods as a substrate for mushroom cultivation not only represents a way of making sustainable use of this agricultural waste, but can also contribute to economic diversification and income generation in the semi-arid region of Paraiba. This practice aligns with the principles of the circular economy, promoting the efficient use of resources and reducing agricultural waste, while offering an opportunity for the development of local production chains and the strengthening of sustainable agriculture in the region.

2.4.5. Algaroba: the algarobeira, or *Prosopis juliflora* (Sw) DC (Figure 09), is a tree that was introduced to the backlands of Pernambuco due to its remarkable adaptation to high temperatures. This species plays a crucial role in local communities, offering a variety of applications ranging from the use of wood for firewood and fencing to providing food for herds during periods of drought, as well as serving as a resource for beekeeping and other uses (Santos *et al.,* 2020).

Figure 09. Leaves of the algarobeira, *Prosopis juliflora* (Sw) DC.

Source: http://shutterstock.com

One of the distinctive characteristics of the algarobeira is its ability to live in symbiosis with rhizobia, which allows it to fix nitrogen in its tissues. The amount of nitrogen fixed in the algarobeira leaf biomass is significant, ranging from 24% to 71% (Dantas, 2022). The composition of algaroba leaves can be seen in Table 03.

Table 03. Composition of the leaves of the algarobeira *Prosopis juliflora* (Sw) DC.

Component	**Content found**
Proteins	14-22%
Crude fiber	21-23%
Calcium	1,5%
Phosphorus	0,2%

Source: Sirmah (2018).

These characteristics make algarobeira leaf tissue a promising option for research into the cultivation of edible mushrooms. By adding this material as a growing substrate, the substrate can be enriched with essential nutrients such as nitrogen and potassium, which can promote the healthy growth and development of mushrooms.

2.5 Nutritional characteristics and application in the food industry

2.5.1 Protein: it is not possible to generalize the protein content of mushrooms, as it depends on several variables, especially the substrate. A protein content of 32% has been reported for *P. ostreatus* grown on sorghum fodder, 32.4% when grown on brewer's grains with wheat bran, and 31.36% on almond and walnut shells and 22.61% of the commercial mushroom of unknown substrate (Majesty *et al.*, 2018; González *et al.*, 2021). The crude protein content of *P. ostreatus* may be higher than in *P. eryngii* and *P. nebrodenis* and with higher protein proportions when grown on olive leaves, followed by grape pomace and wheat straw (Tagkouli *et al.*, 2020). Added to this, it is also reported that in addition to the substrate, there are several strategies to increase the levels of protein and bioactive compounds such as low-dose gamma ray irradiation and UV-B light (Marçal *et al.,* 2021).

2.5.1.a Amino acids: *P. ostreatus* extract is rich in various amino acids (Aas), especially asparaginic acid, histidine, glutamic acid, cysteine and alanine (Dilfy *et al.,* 2020). The most abundant amino acids in *P. ostreatus* is glutamine (7.8-21.69 mg/g), followed by leucine (8.59-16.45 mg/g), alanine (7.22-13.33 mg/g), valine (4.24-9.49 mg/g), glutamic acid (4.81-7.84 mg/g) and serine (3.93-8.83 mg/g) with variations due to the composition of the substrates and strains used. The essential amino acids accounted for 41% of the total amino acids in *P. ostreatus.* And among the essential AAS of the *Pleurotus* species and strains studied recently, leucine predominated, followed by valine, isoleucine, threonine, phenylalanine and lysine, with leucine, valine and lysine predominating in the species grown on various agricultural residues and leucine predominating in the *P. ostreatus* species (Tagkouli *et al.*, 2020).

2.5.1.b. Protein digestibility: the amount of protein accessible for absorption after digestion is determined by the ease with which the peptide bonds are hydrolyzed. And the *in vitro* gastrointestinal digestibility of *P. ostreatus* protein concentrate and flour is satisfactory and supports its high potential for application in a variety of food products, such as protein-fortified foods, vegan foods and specialized nutritional products for people with limited digestive function (Gonzalez *et al.,* 2021).

2.5.1.c. *Recommended Dietary Allowance* (RDA): mushrooms are also considered high-quality nutritional sources due to their ability to meet dietary needs through the Recommended Daily Intake (RDA). Thus, 100 g of mushrooms can provide between 29.41 and 66.0% of the RDA for men and between 35.80 and 80.35% of the RDA for women, while dried meat and whole milk can provide between 47.0 and 59.28% for men and between 57.22 and 67.75% for women (Gonzalez *et al.,* 2020).

2.5.1.d. Protein Efficiency Ratio (PER): this biological method evaluates weight gain as a function of the amount of protein consumed by a growing animal. And although it is in disuse for human nutrition and physiology, some studies have reported that *P. ostreatus* (black oyster mushroom) showed the highest Protein Efficiency Ratio (PER) than other edible mushroom species and higher than that of beef jerky. And other species of the genus showed a PER similar to that of black beans (Mckillop *et al.,* 2021).

2.5.1.d. Protein Efficiency Ratio (PER): this biological method evaluates weight gain as a function of the amount of protein consumed by a growing animal. And although it is in disuse for human nutrition and physiology, some studies have reported that *P. ostreatus* (black oyster mushroom) showed the highest Protein Efficiency Ratio (PER) than other edible mushroom species and higher than that of beef jerky. And other species of the genus showed a PER similar to that of black beans (Mckillop *et al.,* 2021).

2.5.2 Carbohydrates: in *P. ostreatus* they can vary between 46.62% and 50.5% (Bach *et al.*, 2017; González *et al.,* 2021). The majority are polysaccharides that make up the cell wall, such as α- and β-glucans, chitin and hemicelluloses (mannans, xylans and galactans). These polysaccharides are not digestible and are considered a source of dietary fiber (Bach *et al.,* 2017).

2.5.3 Micronutrients: as with other nutrients, the amount of minerals in the basidiomata is also a consequence of the levels of minerals present in the substrates. However, in *P. ostreatus* the ash content varies between 4.1 and 15.9% and is associated with the presence of nutritionally important minerals such as potassium, phosphorus and magnesium. Mushrooms are a rich source of highly and easily absorbable minerals. They have been shown to contain significant amounts of potassium, calcium and sodium. Traces of elements such as iodine, fluorine, copper, zinc, mercury and manganese are also found in the fruiting bodies of mushrooms (Bach *et al.,* 2017). In addition, *P. ostreatus* has high values of vitamin C, vitamin B1, vitamin B12, niacin and folic acid (Dilfy *et al.,* 2020).

2.5.4 Fatty acids: studies agree that *P. ostreatus* contains monounsaturated fatty acids, oleic acid (363 µg/g dried mushroom) and n-6 essential fatty acids, linoleic acid (533 µg/g dried mushroom) in higher concentrations than other mushrooms. However, mushrooms are not considered an important source of essential fatty acids to meet the needs of the human body. Thus, it is possible to infer that *Pleurotus* spp. contain low fat content, but some basic fatty acids (Dilfy *et al.,* 2020).

2.6 Application of mushrooms in the food industry

Consumers are now more interested in using foods that have therapeutic properties, especially after the COVID-19 pandemic. Information to improve health has emerged in recent years, such as the need to reduce consumption of certain foods such as ultra-processed foods and red meat, as they are associated with high levels of cholesterol, saturated fat and other effects that increase comorbidities (Loh *et al.,* 2021). In addition, the harmful consequences of meat on sustainable development mean that an increasing number of people are becoming vegetarians or flexitarians, which means that they want to minimize their meat consumption and choose alternative nutritional sources (Martin *et al.,* 2022; Mora *et al.,*2020). An analysis of 90 dietary recommendations from around the world found that 37% suggested replacing meat protein with plant protein (He *et al.,* 2020). In addition, the market for non-animal protein, which comes from plants and/or edible fungi or insects, will increase (Zozicka *et al.,* 2022; Moura *et al.,* 2023).

To resolve these issues, scientists and the food industry are trying to explore ways of offering alternatives to meat, with the aim of imitating the original animal tissue in terms of texture, aroma, flavor and appearance (Mazumder *et al.,* 2023). The search for more sustainable and healthier alternative protein sources has increased the market for plant-based proteins and popularized their consumption in recent years. However, unlike animal proteins, plant-based proteins are not considered complete protein sources, as cereals and legumes are low in lysine and some lack methionine and cysteine and do not complete all nine essential amino acids (Aschemann-Witzel *et al.*, 2020). But they make up for it due to the lower cost of production when compared to animal production. This leads this niche of consumers to have to vary their sources of cereals to complete their protein supply or supplement.

In this scenario, mushrooms are gaining prominence in the food industry because they have a high protein content, generally above 20g/100g dry weight, and provide all the essential amino acids when compared to plant and animal sources (Ayimbila *et al.*, 2023). In addition, its production for both family farming and industry does not require large areas of land when compared to animal protein and there are economic and environmental gains with the possibility of using agro-industrial waste in cultivation (Aschemann-Witzel *et al.*, 2020).

Studies have already shown the potential of mushrooms as a protein source (Yu *et al.*, 2020, Tagkouli *et al.*, 2020; González *et al.*, 2021,), however, this source is underutilized due to the lack of knowledge about its cultivation, the yields of this crop and its potential as a meat substitute and pharmaceutical compound (Ayimbila *et al.,* 2023). Although the protein value of mushrooms varies depending on the species and strain, the stage of maturity, the substrate used for cultivation and other environmental factors, mushrooms provide significant protein at a lower cost than plant and animal sources (Yu *et al.,* 2020; Kaur *et al.,* 2022).

Mushrooms can be used in the food industry as a future protein alternative, considering the environmentally friendly cultivation system, crop yields, nutritional value, quality, digestibility and biological benefits. One study developed a minced meat substitute based on *Pleurotus spp.* and its results suggested an improvement in textural and sensory attributes with a 37.5% proportion of *Pleurotus sajor cashew* in the compound with chickpea flour, beet extract and canola oil and obtained a product with a protein content of 47% (Mazumder *et al.,* 2023).

Edible mushrooms have recently been used to fortify or enrich foods. According to the Codex General Principles for the Addition of Essential Nutrients to Food, "fortification" is defined as the addition of one or more essential nutrients to a food product, whether or not that nutrient is normally present in the food. Fortification helps to prevent, reduce and control micronutrient deficiencies (Ibrahim *et al.,* 2022). Incorporating mushrooms into food products increases their nutritional value, especially protein and dietary fiber. Pasta enriched with *P. ostreatus* had a higher content of protein, fiber, iron, calcium and potassium (Parvin *et al.,* 2020).

In addition to providing dietary protein, fatty acids, vitamins, fiber and flavor, mushrooms can improve the organoleptic or sensory qualities of processed foods as meat substitutes. Mushrooms are fibrous, which makes meat analogues made with them more chewy (Oosone *et al.,* 2022).

To create an extruded mushroom-based meat substitute with 35% water content, 15% *Lentinus edodes*, *P. ostreatus* and *Coprinus comatus* were used along with isolated soy protein. And the textural profiles of the meat derived from this processing were similar to those of real beef (Yuan *et al.,* 2021). The sweetness and unique flavor of fungi give foods a meat-like taste and improve palatability (Oosone *et al.,* 2020; Mohamad *et al.,* 2020; Ayimbila, 2023).

2.7 Mushroom cultivation in Brazil and worldwide

The global mushroom market is experiencing significant growth, driven by several factors, including increased consumption of healthy foods, diversification of food options and the expansion of the food industry in emerging markets. Global mushroom production is around 40 million tons per year. China remains one of the main leaders in this market with more than half of the world's production with 31.7 million tons produced in 2017, accounting for a substantial part of mushroom production, consumption and exports. Other countries such as the USA (11%), the Netherlands (7%) and several European countries play an important role in global mushroom production. These countries have invested in advanced agricultural technologies and sustainable cultivation practices to increase production efficiency and quality. As such, the forecast that the global mushroom market will reach a value of 69.3 billion dollars by 2024 reflects the potential for continued growth in this sector (Thakur *et al.,* 2020 and Díaz-Godínez *et al.,* 2021).

In Brazil, the adoption of mushrooms in the diet is a recent development and is limited to a few areas where there is a significant concentration of Asian immigrants, as is evident in the state of São Paulo. Despite its potential, the production of edible mushrooms in Brazil is still mostly carried out by small farmers, and is mainly limited to the local market. Approximately 80% of producers are small and medium-sized family farmers who often use traditional, low-tech methods, which contrasts with the progress seen in European and Asian nations (Rodrigues *et al.,* 2022).

It is estimated that there are more than 300 mushroom producers in the country, mainly characterized as micro and small family farmers. The main mushroom species grown and marketed include *Agaricus bisporus* (known as Champignon de Paris), *Pleurotus* spp. (Oyster mushroom), *Lentinula edodes* (Shiitake) and *Agaricus blazei* (Sun mushroom). The states of São Paulo and Paraná stand out as the main leaders in the production of these varieties. There is also significant cultivation in other states such as Minas Gerais, Rio de

Janeiro, Brasília and Rio Grande do Sul. And in the north-east of Brazil, after adaptations and guidance from consultants, there is growing mushroom production in states such as Pernambuco, southern Bahia and Ceará, contributing to the expansion of fungiculture in the region, thus diversifying the national supply of these foods (Rodrigues *et al.,* 2022).

As a result, Brazil has not yet achieved self-sufficiency in mushroom production and depends to some extent on imports to meet domestic demand. In addition, the price of mushrooms on the Brazilian market is remarkably high, largely due to the lack of standardized technologies and the resulting instability in production, among other factors. This is due to challenges in standardizing production, technological limitations, climatic variations and seasonality in supply. As a result, the country still needs to import some of its mushrooms to meet national demand. However, with investments in research, development and the adoption of appropriate technologies, it is possible to move towards self-sufficiency and strengthen national mushroom production (Antunes *et al.,* 2020). In addition, the protein content, availability of essential amino acids and excellent digestibility justify the use of mushroom protein as an alternative protein in the future (Ayimbila *et al.,* 2023).

3 OBJECTIVES

3.1 General Objective

Evaluate the potential of applying low-cost substrates on the productivity and nutritional quality of basidiomata of *P.ostreatus* var. *florida*

3.2 Specific objectives

- To evaluate the influence of light and temperature on the mycelial development of *P. ostreatus* var. *florida,* grown on different grains;
- Selecting formulations based on bean pods without the grain, sawdust, malt bagasse and algaroba leaves for the development of *P. ostreatus* var. *florida*;
- Characterize the individual substrates in terms of pH, reducing, non-reducing and total sugars, ash and soluble solids;
- Carry out the chemical characterization of the formulations: carbon (C), nitrogen (N) and the C/N ratio.
- To classify which of the substrates used is the best formulation for mycelial growth, production and nutritional value of the fungal biomass;
- Characterize the biological parameters: substrate colonization time, primordia emission period, basidiom formation time, basidiom diameter, pileus height and pileus width;
- Characterize the production parameters: yield, biological efficiency, productivity, compost consumption and loss of organic matter of the different formulations;
- Carry out the physicochemical characterization of the mushrooms in terms of total carbohydrates, total proteins, lipids, ash, moisture and total caloric value.

4 MATERIAL AND METHODS

4.1 Places where the **work** was carried out: the experiments were conducted at the Molecular Genetics and Plant Biotechnology Laboratory (L G M Biotec) of the Biotechnology Center - *Campus* I of the Federal University of Paraíba (UFPB), the Plant Tissue Analysis Laboratory of the Agrarian Sciences Center - *Campus* II and the Food Technology Laboratory (LTA/UFPB), *Campus* I, João Pessoa, Paraíba.

4.2. Origin of the mushroom: The matrix of *P. ostreatus* var. *florida* belongs to the fungal culture collection of L G M Biotec/CBIOTEC/UFPB and is being kept under refrigeration at 4 °C in a 2% agar-Sabouraud-dextrose maintenance medium (5 g/L meat peptone, 20 g/L glucose, 5 g/L casein peptone and 15 g/L bacteriological agar), according to the methodology described by Furlan et al. (1997).

4.3 Macro and microscopic examination of structures: the methodology described by Lopes et al. (2008) was used, adapted for microscopic evaluation. The mycelium inoculum was transferred to Petri dishes containing 2% Sabouraud-dextrose agar and covered with a previously flamed coverslip. The plates were kept at room temperature and the coverslips with adhered mycelium were removed successively at 24 - 48 - 72 - 96 - 120 hours after the start of the experiment, placed inverted on a sterile slide containing a drop of lactophenol blue dye and analyzed under an optical microscope to see the vegetative structures. For macroscopic evaluation, a disc of mycelium was inoculated in the center of a Petri dish containing 2% Sabouraud-dextrose agar. The plate was then kept at room temperature (28 ± 2° C). Observations were made for 15 days to characterize the macroscopic appearance of the colony and verify the purity and viability of the culture.

4.4 Adhesive tape technique to capture the spores: the lamellae are arranged on the underside of the basidiomata. A piece of transparent adhesive tape was pressed onto them to capture the spores. The tape was then pressed onto a slide containing a drop of lactophenol blue dye and analyzed under an optical microscope (Arango; Castañeda, 1995).

4.5 Types and origins of substrates: popcorn corn, agricultural gypsum, sawdust, seedless bean pods, algaroba leaves and malt bagasse were used (Figure 10). All the substrates and waste were purchased from local markets in the metropolitan region of João Pessoa, as specified in Table 04.

Figure 10: Substrates and supplements for formulations to test *spawn* production and fruiting of *Pleurotus ostreatus* var. *florida* .

A: Popcorn corn; B: Agricultural gypsum; C: Sawdust; D. Algaroba; E: Green bean pods; F: Malt bagasse: Algaroba; E: Green bean pods; F: Malt bagasse. . Source: Own authorship, 2024.

Table 04. Sources of substrates and supplements.

Materials	**Origin**
Malt pomace	Brewing Technology Laboratory - CBIOTEC/UFPB (João Pessoa - PB)
Green bean pods without the beans	João Pessoa Metropolitan Region Public Market
Popcorn *(Spawn)*	Marketed product
Algaroba leaves	Pernambuco farm in São Mamede/PB
Pine sawdust	Marketed product
Supplements	
Agricultural gypsum ($CaSO_4 .2H_2 O$)	Marketed product

Source: Own authorship, 2024.

4.6 Processing the substrates: all the substrates were treated individually and then mixed to formulate the different treatments as described in Tables 05 and 06.

Rice in the husk (*Oryza sativa* L.): used as the base substrate only for evaluating lighting and temperature. The grain was washed with running water to remove excess starch, pre-cooked with distilled water and agricultural gypsum for 8 min in a microwave at high power and then autoclaved at 121 °C for 60 minutes (Methodology adapted from Salami *et al.*, 2016).

Malt bagasse (*Hordeum vulgare* L.): this waste came from the processing of beer and, due to its high humidity, did not undergo any pre-treatment. After collection, it was packed in plastic bags and kept refrigerated at 4 C.°

Unshelled green bean pods (*Phaseolus vulgaris* L.): after collection, the green bean pods were placed in a drying oven for 72 hours at 60 ± 5° C. After drying, they were submerged in running water for 24 hours. The water was then drained off and the pods were left on a solid platform to dry externally for approximately 2 hours at room temperature (28 ± 2° C). After this, the pods were crushed in a milling machine and then packed in plastic bags and kept refrigerated at 4° C (Figure 11) (Zanetti; Ranal, 1997).

Figure 11. Green bean pod without grain, after processing.

A: Pods submerged after drying. B: Draining the pods. C: Damp material. D: Crushed pods. Source: Own authorship, 2024.

Popcorn (*Zea mays* var. *everta*): the grain was cooked over high heat in a pressure cooker for 90 minutes. After cooking, the extract from the cooking process was drained off and the grain was washed with water and left at room temperature (28 ± 2° C) to dry externally for approximately 60 minutes. They were then placed in plastic bags and kept refrigerated at 4° C (Bonatti *et al.*, 2004).

Leaves of the algaroba tree (*Prosopis juliflora* (Sw) DC): after collection, the leaves were placed in a drying oven for 72 hours at 60 ± 5° C. After drying, the leaves were submerged in running water for 24 hours. The leaves were then pasteurized by immersing them in boiling water for 3 hours to remove easily soluble sugars and contaminants. Afterwards, the water was drained off and the leaves were left on a solid platform to dry externally for approximately 2 hours at room temperature (28 ± 2° C). They were then packed in plastic bags and kept refrigerated at 4° C until use (Fermor; Wood, 1991).

Agricultural gypsum ($CaSO_4 .2H_2 O$): was used following the manufacturer's recommendations without undergoing any dilution process. As well as retaining moisture, calcium sulphate provides Ca and S and is absorbed by the grains and then used by the mushroom mycelium.

Sawdust (*Pinus elliottii*): material collected from animal houses and used after being submerged in running water for 24 hours, after which the water was drained and packed in plastic bags and kept refrigerated at 4°C.

Table 05. Formulations tested for the production of *spawn* and basidiom of *Pleurotus ostreatus* var. *florida* for small-scale production.

Treatments/Acronyms	Formulations/Substrates
TC*	Corn (100%)
T1 (BM)	Malt Bagasse (100%)
T2 (VF)	Green bean pods without beans (100%)
T3 (SE+GA)	Sawdust (100%) + Agricultural Gypsum (2%)
T4 (BM+GA)	Malt Bagasse (100%) + Agricultural Gypsum (2%)
T5 (VA+GA)	Green Bean pods without grains (100%) + Agricultural Gypsum (2%)
T6 (FA+GA)	Algaroba Leaves (100%) + Agricultural Gypsum (2%)
T7 (BM+VF)	Malt Bagasse (50%) + Green Bean Pods (50%)
T8 (BM+FA)	Malt Bagasse (50%) + Algaroba Leaves (50%)
T9 (BM+SE)	Malt Bagasse (50%) + Sawdust (50%)
T10 (BM+VF+GA)	Malt Bagasse (50%) + Green Bean Pods without Grains (50%) + Agricultural Gypsum (2%)
T11 (BM+FA+GA)	Malt Bagasse (50%) + Algaroba Leaves (50%) + Agricultural Gypsum (2%)
T12 (BM+SE+GA)	Malt Bagasse (50%) + Sawdust (50%) + Agricultural Gypsum (2%)
T13 (VF+FA)	Green Bean pods without beans (50%) + Algaroba leaves (50%)
T14 (VF+FA+GA)	Green Bean Pods without Grains (50%) + Algaroba Leaves (50%) + Agricultural Gypsum (2%)

TC: Control treatment for *spawn* production test only. Source: Own authorship, 2024.

All the formulations were mixed manually and then autoclaved in a vertical autoclave (Prismatec - CS) at 1 atm for 60 min. at 121 °C before inoculation.

4.7 *Spawn* production: the inoculum or "*spawn*" was prepared according to the methodology described by Bonatti *et al.* (2004), consisting of popcorn kernels used as a support for mycelial growth. 50g of cooked popcorn kernels were mixed with agricultural gypsum (2% w/w) and placed in glass jars (6.0 x 8.5 cm). The jars were sealed with aluminum foil to facilitate gas exchange and sterilized in an autoclave at 121° C for 60 minutes. After sterilization, each jar was inoculated with three agar discs (10 ± 1 mm in diameter) containing fungal mycelium and incubated at 28° ± 2°C in the dark until the mycelium had completely colonized the surface of the grains (Figure 12).

Figure 12. Seed inoculum of *Pleurotus ostreatus* var. *florida* on popcorn kernels.

A: Pre-cooked and autoclaved popcorn and B: Popcorn colonized with the primary mycelium of *P. ostreatus* var. *florida*. Source: Own authorship, 2024.

4.8. Evaluation of the influence of light on mycelial growth: treatments TLT1 to TLT3 with base formulations (Table 06) were used. The experiment was carried out in test tubes (2.5 x 20 cm). After inoculation with 2% of the *spawn*, the tubes were kept in intermittent light, intermittent dark and a 12 h light/12 h dark photoperiod at room temperature. The experiment was carried out in triplicate and evaluated over 15 days. Measurements were taken vertically with a millimeter ruler (cm).

4.9. Evaluation of the influence of temperature on mycelial growth: treatments TLT1, TLT2 and TLT3 were also used (Table 06). The experiment was carried out in test tubes (2.5 x 20 cm) containing formulations already known to support primary mycelial growth. After inoculation with 2% of the *spawn*, the tubes were kept at 20° C, 25° C, 30° C and 35° C in a bacteriological oven in the dark. The experiment was carried out in triplicate and evaluated over 15 days. The measurement was made vertically with a millimeter ruler (cm) (Imtiaj *et al.,* 2008).

Table 06. Formulations for analyzing the influence of light and temperature on the mycelial run of *Pleurotus ostreatus* var. *florida*

Treatments	Substrate formulations/proportions
TLT1	Rice (98%) + Agricultural Gypsum (2%)
TLT2	Malt Bagasse (100%)
TLT3	Malt Bagasse (96%) + Agricultural Gypsum (2%)

TLT: Light and Temperature Treatment. Source: Own authorship, 2024.

4.10 Selection of treatments/substrates by evaluating vigor, density and mycelial growth for the production of *spawn* and basidiom of *Pleurotus ostreatus* var. *florida*: treatments TC to T14 were used (Table 05). The substrates were weighed (10 g), mixed manually and placed in glass jars with screw caps (4.0 x 10 cm) and then autoclaved at 121 °C for 30 minutes. After sterilization, 2% of the *spawn* produced on the popcorn was inoculated. The jars were then incubated in the dark at room temperature. The speed at which the mushroom colonized the different formulations was assessed by measuring the growth of the mycelium (in mm) using a millimeter ruler, considering the top of the substrate as point 0 (zero) (Pedra; Marino, 2006). In this experiment, the speed of mycelial growth of the mushroom itself was not taken into account, but rather the average growth (cm/day) of the mushroom as a function of each substrate. To assess vigour and density, the mycelium was classified as: strong mycelium (characterized by high density, compactness and strong vigour); medium mycelium (characterized by medium density, little compactness and less vigour) and; weak mycelium (characterized by low density, sparse and no vigour) (Bernardi *et al*., 2013). For all the evaluations, an experiment was carried out in triplicate (Figure 13).

4.11. Effect of substrates/treatments on growth time for *spawn* production of *Pleurotus ostreatus* var. *florida*: the substrates/treatments selected on the basis of vigor, density and mycelial growth were used in this test. The substrates were weighed (60 g), mixed manually and packed in polypropylene bags (6 cm x 25 cm), sealed with nylon ties and then autoclaved at 121°C for 30 minutes. After sterilization, 2 % of the *spawn* was inoculated. The bags were then incubated in the dark at room temperature. The speed of colonization of the mushroom in the different formulations was assessed by measuring the growth of the mycelium (in mm) using a millimeter ruler, considering the top of the substrate as point 0 (zero) (Pedra; Marino, 2006). A test was carried out with three replicates.

4.12. *Pleurotus ostreatus* var. *florida* production trial: the axenic cultivation technique was used, in which the treatments/substrates are sterilized and enriched with nutrients, allowing the fungus to grow without competition and obtain greater productivity (Pedra; marino, 2006). Based on the characterization of mycelial growth, density and vigor, physicochemical analyses were carried out on the best treatments to verify the influence of the formulations on the nutritional content of the basidiomata. 100 g of the substrate/treatment were placed in bags (12 cm x 20 cm). After sterilization, the bags were inoculated with 2% of the *spawn*. The bags were then incubated in the dark and at room temperature ranging from 25 to 32°C for 15 days and then opened for the fruiting phase. After opening the bags, all the treatments were subjected to daily watering with tap water, manually with sprayers. After fruiting, the basidiomata were harvested when they showed maximum development, as seen by the beginning of the unwinding of the pileus margins. An experiment with five replicates was carried out for each treatment.

4.12.1 Evaluation of biological parameters: during the cultivation period, biological aspects of the mushroom were analyzed, such as: time of colonization of the substrate, period of emission of primordia, time of formation of basidiocarps, height of the stipe (cm) and diameter of the pileus (cm).

4.12.2 Evaluation of production parameters: during the cultivation period, mushroom production parameters were analyzed according to the cultivation substrate, such as: Mushroom and treatment humidity (U%), Yield (R%), Biological Efficiency (EB%), Productivity (Pr g/day^{-1}) and Compost Consumption (CC%).

4.12.2.1 Moisture of the substrates, formulations and mushrooms: moisture was (υ%) was determined by the gravimetric method using heat, which is based on the weight loss of the material when heated until it reaches a constant weight. All the formulations were kept in a drying oven at 105° C for 24 hours and the moisture content of the substrate was determined by the difference between the wet and dry biomass according to Equation 01 (AOAC, 1995).

$$U\ (\%) = \frac{(MIA\ -\ MFA) * 100}{MIA}$$

Eq. 01

In which:

υ - Humidity

MIA - Initial sample mass

MFA - Final sample mass

Yield: to determine the yield (R %) of the process, the relationship proposed by Chang *et al.* (1981) was used, which relates the wet mass of the fruiting bodies to the dry mass of substrate (Equation 02).

$$R\,(\%) = \frac{MCU \times 100}{MSSI}$$

Eq. 02

In which:

MCU = Wet mass of fruiting bodies, g

MSSI = Dry mass of initial substrate, g

Productivity: the productivity (Pr g/day^{-1}) of the process was determined according to Holtz (2009). It consists of the ratio between the mass of dried fruiting bodies and the total cultivation time (time from inoculation to production flow) (Equation 03).

$$Pr[g.dia^{-1}] = \frac{MCS}{TC}.$$

Eq. 03

In which:

MCS = Dry mass of fruiting bodies, g

TC = Total cultivation time, days

Biological Efficiency (EB%): the biological efficiency (EB%) of the process was determined by the ratio between the mass of fruiting bodies and the mass of substrate based on their dry weight according to Equation 04 (Bisaria et al., 1987).

$$EB(\%) = \frac{MBS}{MSS} \times 100$$

Eq. 04

MBS = Mass of dried basidiomata, g

MSS = Dry mass of substrates, g

4.13 Chemical analysis of **the raw material:** malt bagasse, bean pods without grains and algaroba leaves were analyzed individually using 3 replicates of each substrate for the analysis of reducing sugars, non-reducing sugars, total sugars, minerals, pH and soluble solids (Brix°).

Sample preparation for analysis: All the substrates were ground in a knife mill, dried in a LUCA-80/27 oven at 102°C for 24 hours and stored at room temperature between 25 and 32°C.

pH: The pH of the substrates was determined using a bench pH meter previously calibrated with a buffer of 7 and 4. 5g of each dry substrate had to be suspended in 50 mL of distilled water. The suspension was homogenized using a magnetic stirrer.

Ash: Minerals were determined by carbonizing 5 g of the material in triplicate in a muffle furnace at 550°C to decompose all the organic parts and the results were expressed in % according to the methodology of Nogueira and Souza (2005).

Reducing sugars: 10g of the sample was diluted and filtered with 250mL of distilled water, 25ml of the filtrate was transferred to a burette and 10ml of Fehling's solutions A (copper sulphate) and B (sodium potassium tartrate and sodium hydroxide) were pipetted and transferred to a 250ml Erlenmeyer flask topped up with distilled water and heated to boiling point which was kept constant during the titration until a brick red color appeared. Methylene blue 1% was used as an indicator. Sugars were measured as a percentage using Equation 05:

$$\%Açúcares\ redutores = \frac{F \times C \times 100}{P \times V}$$

Eq. 05

F = Fehling's liquor factor (F/2 when using 5 ml)

C = volumetric capacity of the flask used (250 ml)

P = sample weight

V = volume spent on titration

Solutions with a pH below 6 were neutralized with 0.1N NaOH. And samples with intense colors were clarified with 1.0M zinc acetate and 10 ml of 0.25M potassium ferrocyanide.

Total sugars: Total sugars were determined using 5 g of the sample diluted and filtered with 250 ml of distilled water, 10 ml of concentrated HCL was added and the solution placed in a water bath at 68 to 78°C for 20 minutes. After cooling, the solution obtained was transferred to a 25 ml burette, 5 ml of Fehling's solutions A and B were volumetrically pipetted into the solution, distilled water was added and heated to boiling. The sample was titrated while boiling until a brick-red color appeared, using 1% methylene blue as an indicator as per the Lane-Eynon method (2008). Total sugars were calculated using Equation 06:

$$\%\,Açúcares\ totais\ = \frac{(F \times C \times 100)}{(P \times V)}$$

Eq. 06

In which:

F = Fehling liquor factor

C = volumetric capacity of the flask used - 250 mL

V = volume spent on titration

Non-reducing sugars: These are classified as such due to the absence of free aldehyde and ketone groups. They are not capable of reducing salts and need to undergo hydrolysis of the glycosidic bond in order to oxidize. Non-reducing sugars were calculated using the difference between total sugars and reducing sugars according to Equation 07.

$$\%Açúcares\ não\ redutores\ = Açúcares\ totais\ - \ Açúcares\ redutores$$

Eq. 07

Soluble solids (°brix): a 2g sample of each dry substrate was diluted with 15ml of distilled water and homogenized. The large particles were then discarded. Afterwards, 1 to 2 drops were transferred to the prism of the RHB0-90 analog refractometer for % Brix and read on the scales in °Brix.

Chemical analysis of the mushrooms: 3 replicates of each treatment were used for the chemical analysis of the fruiting bodies. The mushrooms were weighed, dehydrated at 55 °C for approximately 36 hours, ground in a mini-processor, packed in polyethylene jars, closed and stored under refrigeration at 5 ± 2 °C. The following determinations were made on the basidiocarps, according to the procedures adopted by the A.O.A.C (1984) for moisture, lipid content and ash. Moisture was quantified in an oven at 105 °C for 6 hours. Crude fat content was determined by gravimetry after continuous extraction of the samples with sulfuric ether in Soxhlet equipment and ash was calculated by the mass of the sample after incineration. The protein fraction was determined using the "Kjeldahl" method, with the crude protein of the mushroom being determined from the nitrogen content, using the conversion factor N x 4.38 according to Miles and Chang (1997).

Carbohydrates and caloric value: estimated by difference from moisture, fixed mineral residue, protein and lipids expressed in g/100g, using the following calculation (Equations 08 and 09) (INSTITUTO ADOLFO LUTZ, 1985):

$$\%\, Carboidratos = 100 - (A + B + C + D)$$

Eq. 08

In which:

A - % humidity

B - % fixed mineral residue

C - % protein

D - % lipids

Total caloric value: was calculated according to the calculation (ANVISA, 2003):

$$Valor\ Calórico\ Total\ (Kcal/100g) = (E \times 4) + (C \times 4) + (D \times 9)$$

Eq. 09

In which:

E - carbohydrates

C - proteins

D - lipids

Total **Proteins**: Total proteins were determined using the Kjedahl method according to the Manual of Food Analysis Practices (2020), which is based on wet combustion via heating with concentrated sulfuric acid in the presence of catalysts to reduce the organic nitrogen, ammonia, captured in an alkaline solution to form ammonium sulfate. The ammonia is then distilled off in a 4% boric acid solution followed by direct titration of the ammonia with a standard hydrochloric acid solution. The sample digestion procedure was carried out by carefully transferring 1.0g of the samples to the Kjeldahl tube with 1.8g of catalytic mixture and 10mL of concentrated sulphuric acid without touching the tube walls. The tubes were placed in the digester at 350°C until the solution became colorless or slightly bluish. After cooling the tube, the walls were washed with 5mL of distilled water and then 3 drops of phenolphthalein were added. It was then attached to the distiller. The condenser outlet was immersed in an erlenmeyer flask with 25 mL of 4% boric acid H_3BO_3 solution. The boiler water was previously heated and then the distiller temperature was lowered to a scale of 3. The alkaline pH was obtained by adding 40% NaOh until the color changed to purple. The temperature of the distiller was then increased to 10 in order to promote distillation. The boric acid solution containing the distilled protein was reserved for titration with 0.1M hydrochloric acid until a reddish color appeared. The results were expressed as % nitrogen and the amount of protein was determined using Equation 12.

$$Proteínas\ totais(g/100g) = \frac{(Va - Vb)x\ fa\ x\ F\ x\ 0{,}14}{Peso\ da\ mostra}$$

Eq. 10

Va - volume of standardized 0.1N hydrochloric acid used to titrate the sample

Vb - volume of standardized 0.1N hydrochloric acid used in the blank titration

fa - nitrogen - vegetable protein matching factor: 5.75

Total lipids: total lipids were determined using the Bligh & Dyer method (1959) in which 2.5 g of the dried samples in triplicate were subjected to cold extraction of the total lipids in a mixture of chloroform, methanol and water. In this method, the lipids remain in the chloroform phase, which is then evaporated (INSTITUTO ADOLFO LUTZ, 2008). The amount of lipids is obtained by weighing and the results are expressed in g per 100g of sample using Equation 13:

$$Lipídios\ totais\ (\%) = \frac{peso\ final - peso\ béquer\ x\ 4x\ 100}{Peso\ da\ mostra\ (g)}$$

Eq. 11

4.14. N, C and C/N content

Organic carbon: determined using the Walkley and Black method, modified from Tedesco et al. (1995). Each sample (0.01 g) was added to a flask containing potassium dichromate ($K_2Cr_2O_7$) and sulfuric acid ($H_2\ SO_4$) and heated to 60°C. Two types of white were prepared: one that was subjected **to** the same heating conditions and one that was not heated. After cooling, indicators were added and then titration was carried out with a solution of ammoniacal ferrous sulphate ($Fe(NH)_{42}(SO)_{42}.6H_2O$). The results were obtained from Equations 10 and 11:

$$A = \frac{((Vba - Vbn)(Vbn - Vba))}{Vbn + (Vba - Vam)}$$

Eq. 12

In which:

Vba: volume spent titrating the blank with heating;

Vbn: volume spent titrating the blank without heating;

Vam: volume spent titrating the sample.

$$Carbono\ (C)\% = \frac{(A)(molaridade\ do\ sulfato\ ferroso)(3)\ x\ 100}{(massa\ da\ amostra\ (mg))}$$

Eq. 13

In which:

3= ratio between the number of moles of dichromate reacting with iron, multiplied by the number of moles of dichromate reacting with carbon, multiplied by the atomic mass of carbon.

Total nitrogen: the Kjeldahl method was used to analyze total nitrogen, where the samples (0.1g) were digested in a catalyst mixture containing sulfuric acid (H_2SO_4) in a Velp Scientifica digester block, until a blue color was obtained. After cooling, the reactions were distilled in a Velp Scientifica nitrogen distiller in a basic medium (NaOH 35%) and the product of this distillation was collected in an erlenmeyer flask in a 4% boric acid solution (H_3BO_3) with indicators and titrated with 0.1 N hydrochloric acid (HCl) (Tedesco et al., 1995, Moretto et al., 2002a). The nitrogen concentration was determined according to Equation 12:

$$N(N)\% = \frac{V\ x\ Fc\ x\ 0{,}0014}{M}$$

Eq. 14

In which:

V = volume of HCl used in the titration, mL.

Fc= HCl correction factor

M = sample mass (g)

C/N ratio: The C/N ratio was obtained from the proportion of carbon contained in each sample in relation to nitrogen.

4.15. Statistical analysis: the experimental design was complete randomized blocks, using the Tukey test, with a significance level of $p<0.05$ (GOMES, 1990). Statistical analysis was carried out using the SAS (Statistical Analysis System) program. The results were submitted to analysis of variance (ANOVA) followed by the Tukey test with a significance level of 5%.

5 RESULTS AND DISCUSSION

5.1 Micro and macroscopic **analyses of the vegetative and reproductive phases of *Pleurotus ostreatus* var. *florida*:** the macroscopic appearance of the *P. ostreatus* colony can be seen in Figure 14.A. It is characterized by a fibrous texture that is completely white on the front and back of the colony. After inoculation, the mushroom showed rapid radial growth in approximately 9 days on Agar-Sabouraud-dextrose culture medium. The medium used proved to be an ideal solid nutrient medium for preserving and evaluating macro and microscopic characteristics and for vegetative propagation *in vitro*.

Through microscopic evaluation, it was possible to observe the basic structure of the fungus - uninucleate septate hyaline hyphae that make up the primary mycelium (Figure 13.B. 1 and 3) - characterized by connecting clamps or anastomoses (Figure 13.B. 2) through which nuclei migrate between the mycelia. In an appropriate culture medium, this dicary forms the fruiting body by the accumulation of hyphae, resulting in a macroscopic structure. The hyphae generally have an apical growth through the accumulation of vesicles at the apex of the hyphae (Figure 13. C) through which lignocellulolytic enzymes are excreted to advance the primary mycelium into the substrate (Figure 13. D). Usually, when the fungus has completely consumed the substrate, together with environmental variations such as increased humidity and lighting, the hyphae undergo cell differentiation to form secondary mycelium (dicari) and the fruiting body (Figure 13. E) which, when mature, is characterized by the strain and pileus with a color ranging from white to light grey (Figure 13. F). The lamellae are arranged on the underside of the basidiome (Figure 13. G) and the basidia are located inside, where meiosis takes place. Each basidium produces four basiodiospores, smooth, sub-cylindrical sexual spores (Figure 13.H). The spores in a suitable medium can germinate and repeat the cycle.

Figure 13. Morphological analysis of *Pleurotus ostreatus* var. *florida* and biological cycle.

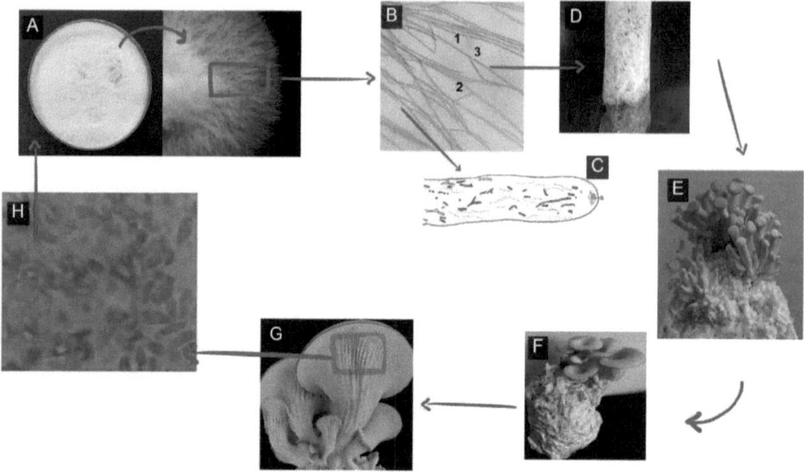

A: Petri dish with vegetative mycelium on Sabouraud-dextrose agar medium. B: Connecting staples (1. hypha, 2. anastomosis, 3. septum); C: Cellular characteristic of the hypha with arrangement of organelles and apical growth zone S. D: Growth of the vegetative mycelium on a lignocellulosic substrate composed of malt bagasse. E: Differentiation of the somatic phase into primordia of basidiomata. F: Mature fruiting body (basidiome). G: Lamellae. H: Spores. Source: Own authorship, 2024.

5.4 Influence of light on mycelial growth: The lowest growth occurred in 24 hours of light with TLT1 having the highest mycelial growth (2.1 cm), followed by treatments TLT2 (1.2 cm) and TLT3 (0.97 cm). During the photoperiod, treatment TLT1 (3.25 cm) achieved the greatest growth, followed by TLT3 (2.8 cm) and TLT2 (2.7 cm). After 24 hours of listening, treatment TLT1 (3.5) also achieved the greatest mycelial growth, followed by TLT2 (2.96 cm) and TLT3 (2.7 cm). Greater growth was also observed in the treatments composed of Rice (98%) and Agricultural Gypsum (2%) (TLT1) in the different photoperiods (Figure 14).

Figure 14: Effect of lighting on the growth of the primary mycelium of *Pleurotus ostreatus* var. *florida* in different environmental conditions.

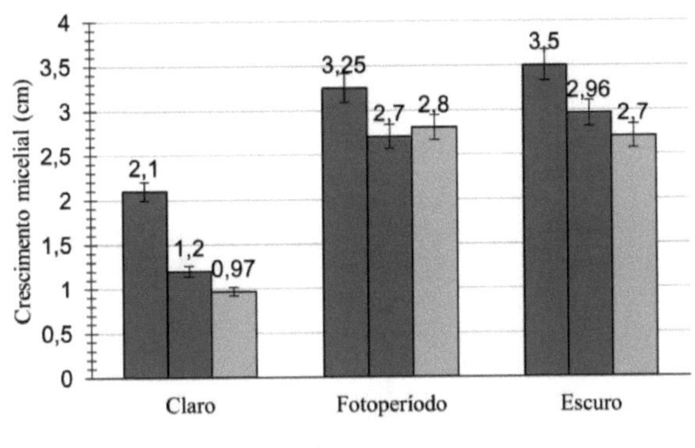

TLT1 [Rice (98%) + Agricultural Gypsum (2%)]; TLT2 [Malt Bagasse (100%)] and TLT3 [Malt Bagasse (96%) + Agricultural Gypsum (2%)]. Source: Own authorship, 2024.

Thus, the mycelial growth of *P. ostreatus* var. *florida* was better evidenced in the dark and in the photoperiod. It was more stable in the dark in all treatments (Figures 14 and 15). The test kept in the light showed slower growth and less vigor when compared to the treatments kept in the dark and the 12 h light/12 h dark photoperiod. The behavior of the fungus in the vegetative phase is an important factor for productivity and rapid growth, vigor and mycelium density should be assessed. Thus, according to the results, direct lighting can slow down mycelial growth as well as seed-inoculum production. The results show a pattern found in other fungi, as reported by Zheng et al. (2021). At 24 hours of illumination, there was a repression of mycelial growth in filamentous fungi such as *Botrytis cinerea, Aspergillus fumigatus and Cordyceps militaris*.

In a similar study under different photoperiod extensions, the mycelial growth of *B. ochroleuca* showed a similar pattern in response to temperature change, but was inhibited by light, which caused a 10% to 20% decrease in growth at all temperatures. Extending the duration of light significantly reduced the mycelial growth of the fungus (Zheng *et al.,* 2021). This demonstrates the strong influence of photoperiod only on the reproductive phase of some species. And in the case of agaricomycetes, it is crucial for the development of basidiomata. This means that photoperiod does not have a strong influence on the vegetative growth of mushroom species. It does not affect artisanal production with natural lighting.

In this study, all the treatments that spent 24 hours in the light had a delay in mycelial growth when compared to the treatments submitted to dark and photoperiodic conditions, which were ideal for the development of the primary colonization phase of *P. ostreatus* var. *florida.*

Figure 15: Effect of light on the colonization of *Pleurotus ostreatus* var. *florida* in different environmental conditions.

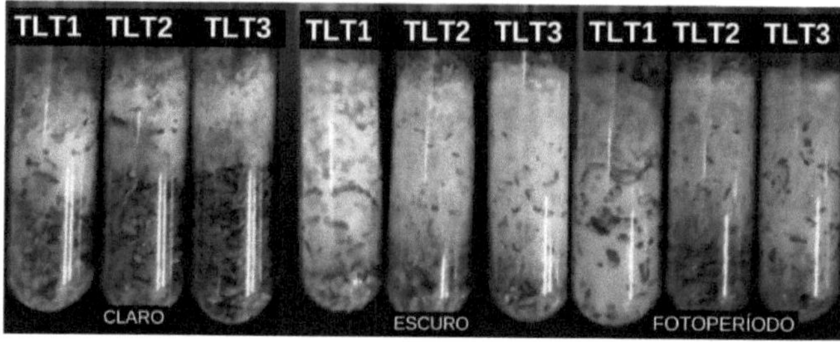

TLT1 [Rice (98%) + Agricultural Gypsum (2%)]; TLT2 [Malt Bagasse (100%)] and TLT3 [Malt Bagasse (96%) + Agricultural Gypsum (2%)]. Source: Own authorship, 2024.

White light, whether artificially emitted or not, can be associated with stress factors for *P. ostreatus.* Lighting can be associated with heat, leading the organism to osmotic stress and greater energy expenditure in order to continue producing lignocellulolytic enzymes (Schumacher *et al.*, 2017). Despite not being a photosynthetic organism, lighting can provide signals about the environment to the fungus, which is necessary for its survival. The results obtained in the present study are in line with tests by Markson *et al.* (2012), who obtained a larger colony diameter in completely dark environments. However, other investigations into the effect of light on the mycelium of basidiomycetes have obtained satisfactory growth of *L. sajor-caju and S. commune* in both lit and unlit conditions. Thus, it is also inferred that the effect of light on the primary development of basidiomycetes can vary between species (Alfonso *et al.*, 2016).

5.5 Evaluation of the influence of temperature on mycelial growth: studies on the effect of ambient temperature are necessary for the correct environment of a fungal strain for artisanal mushroom cultivation in a given location. In this study, the vigor and density of the primary mycelium of *P. ostreatus* var. *florida* varied according to the temperatures tested (20°C, 25°C and 30°C) in the different treatments. At 20°C there was the greatest growth in the TLT2 treatment (6.9 cm), followed by TLT3 (6 cm) and TLT1 (5 cm). At 25°C all treatments reached maximum mycelial growth (7 cm). At 30°C the TLT2 treatment (5.7 cm) achieved the greatest growth, followed by TLT1 (5.07) and TLT3 (4.97 cm) (Figure 16).

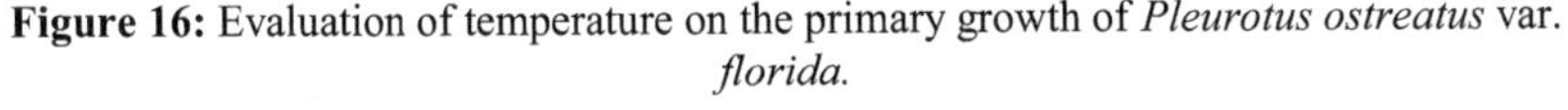

Figure 16: Evaluation of temperature on the primary growth of *Pleurotus ostreatus* var. *florida.*

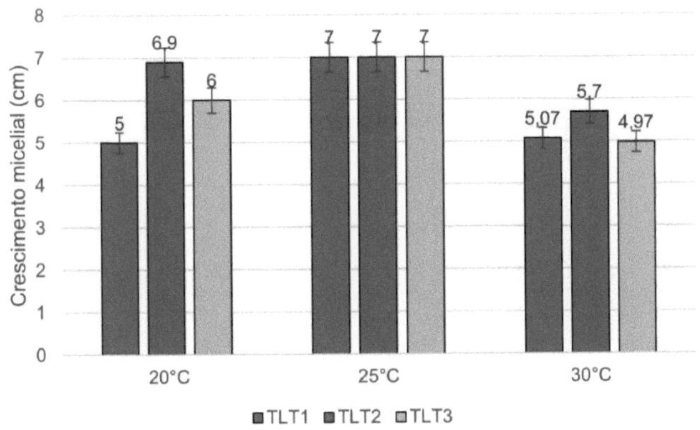

TLT1 [Rice (98%) + Agricultural Gypsum (2%)]; TLT2 [Malt Bagasse (100%)] and TLT3 [Malt Bagasse (96%) + Agricultural Gypsum (2%). Source: Own authorship, 2024.

From the tests at the different temperatures, it was not possible to see any significant difference in mycelial growth between the different substrate groups (Figure 07). However, with the increase in temperature to 30°C, colony growth was slower and mycelial vigor decreased noticeably, showing a less dense and compact appearance of the mycelium when compared to the other groups.

The results shown in Figures 16 and 17 are in agreement with other studies which have shown that the *spawn* production time of *Pleurotus* sp. decreases at temperatures from 12 to 20°C and has colonization optimization at temperatures close to 25°C (Kibar; Pekien, 2008; Charan *et al.,* 2022). In another similar investigation, eight environmental temperatures (15°C, 18°C, 20°C, 22°C, 25°C, 28°C, 30°C and 32°C) were used to grow *P. ostreatus* mycelia on cottonseed hulls as a substrate. Mycelial growth and fruiting differed significantly between temperatures, with 22°C being ideal for the *P. ostreatus* strain. However, the highest rate of mycelial growth occurred at a temperature of 28°C, but at this temperature it took longer for primordia to form (Hu *et al.,* 2023). There may be an increase in the activity of laccase, an early lignocellulose-degrading enzyme, at temperatures between 18-22°C and, with increasing temperature, higher activities of proteinases, responsible for the assimilation of nitrogen involved in the late phase of vegetative growth aimed at the cellular differentiation of primary mycelium into secondary mycelium to form basidiomata (Jiang *et al.*, 2020). This indicates that a subtly lower temperature for *spawn* production would be more suitable for this stage of development.

Figure 17: Qualitative assessment of the mycelial growth of *Pleurotus ostreatus* var. *florida.*

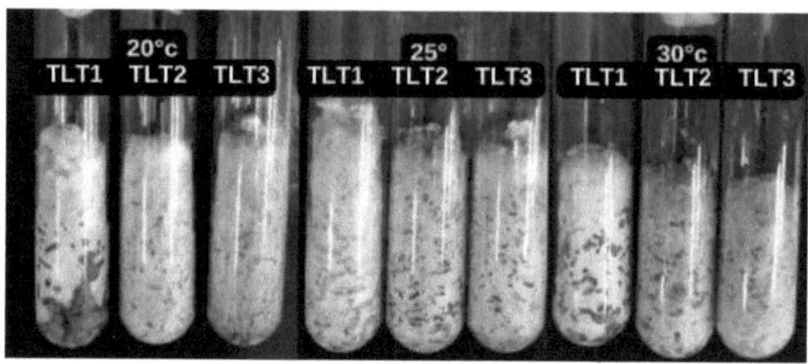

TLT1 [Rice (98%) + Agricultural Gypsum (2%)]; TLT2 [Malt Bagasse (100%)] and TLT3 [Malt Bagasse (96%) + Agricultural Gypsum (2%) + Coffee grounds (2%)]. Source: Own authorship, 2024.

Although the intensity of mycelial growth, which is fundamental to the development of edible mushrooms, varies significantly according to the cultivation temperature of *P. ostreatus*, when carrying out a comprehensive analysis of the rate of mycelial growth and the vigor of development in these trials, it is clear that the optimum temperature range for the culture of the mycelium is between 20 and 25°C. However, it is important to note that discrepancies can arise due to the use of different culture media and the peculiarities of species and strains.

5.6 Selection of substrates and formulations by evaluating vigor, density and mycelial growth for the production of *spawn* and basidiom of *Pleurotus ostreatus* var. *florida*: among the 14 treatments tested (Figure 18), all the formulations and substrates tested contributed to the qualitative growth of *P. ostreatus*, with the exception of treatments composed of pine sawdust such as T3 (SE+GA), T9 (BM+SE) and T12 (BM+SE+GA). They showed low mycelium density, vigor and slow growth. It is possible that the resin content and extractive fraction (phenolic materials such as lignans, flavonoids, phenols and polyphenols) that play a protective role in plant structure found in pine wood may have inhibited or slowed down the growth of *P. ostreatus* (Rugolo *et al.*, 2020). Therefore, the formulations consisting of sawdust (T3, T9 and 12) were removed from the following stages.

Figure 18. Qualitative assessment of the mycelial growth of *Pleurotus ostreatus* var. *florida* in the 15 treatments.

TC (Boiled maize (100%)); T1(Malt bagasse (100%); T2 (Green bean pods without grains (100%); T3 (Sawdust (100%) + Agricultural gypsum (2%); T4 (Malt bagasse (100%) + Agricultural gypsum (2%); T5 (Bean pods without grains (100%) + Agricultural gypsum (2%); T6 (Algaroba leaves (100%) + Agricultural gypsum (2%) T7 (Malt bagasse (50%) + Bean pods without grains (50%); T8 (Malt bagasse (50%) + Algaroba leaves (50%); T9 (Malt bagasse (50%) + Sawdust (50%); T10 (Malt bagasse (50%) + Green bean pods without grains (50%) + Agricultural gypsum (2%); T11 (Malt bagasse (50%) + Algaroba leaves (50%) + Agricultural gypsum (2%); T12 (Malt bagasse (50%) + Sawdust (50%) + Agricultural gypsum (2%); T13 (Green bean pods without grains (50%) + Algaroba leaves (50%) and T14 (Green bean pods without grains (50%) + Algaroba leaves (50%) + Agricultural gypsum (2%). Source: Own authorship (2024).

Based on these results, the treatments selected for the *Pleurotus ostreatus* var. *florida spawn* production test were T1, T2, T4, T5, T6, T7, T8, T10, T11, T13 and T14. All these treatments contained bean pods without the grain, algaroba leaves and malt bagasse. The physicochemical characterization of these substrates was carried out.

5.7 Physicochemical characterization of the selected substrates: the physicochemical analyses of the isolated substrates showed that malt bagasse (BM) contained 4.5% total sugars, 2.68% reducing sugars and 1.83% non-reducing sugars. On the other hand, bean pods without grains (VF) contained 3.62% total sugars, 2.17% reducing sugars and 1.34% non-reducing sugars. It was not possible to quantify the total sugar content of algaroba leaves (AF) with the methodology used. The BM showed 3.41% ash, 8.09% moisture, pH 5.94 and 1° Brix. The VF showed 2.24% minerals, 22.10% moisture, pH 6.30 and 1° Brix and the FA showed 5.55% minerals, 6.9% moisture, pH 5.35 and 1° Brix (Table 07).

Table 07: Composition of the raw material.

Substrates	Total sugars	Reducing sugars	Non-reducing sugars	Minerals	Humidity	pH	Brix°
Malt Bagasse (BM)	4,5±0,011	2,68±0,009	1,83±0,47	3,41±0,04	8,09±0,14	5,94	1°
Bean pods without beans (VF)	3,62±0,09	2,17±0,021	1,34±0,076	2,24±0,33	22,10±21,52	6,30	1°
Algaroba leaves (FA)	N/A	N/A	N/A	5,55±0,36	6,91±0,09	5,35	0,1°
CV (%)	11,08	26	11,99	39,54	11,33		

CV: Coefficient of Variation (%). Source: Own authorship, 2024.

In mushroom cultivation, substrates must contain the carbon sources that provide the structural element of fungal cells in combination with other essential elements such as hydrogen, oxygen and nitrogen. The assimilation of glucose by fungi leads to their respiration, the metabolic form of energy production initiated by glycolysis. Thus, many carbon-rich substrates can be used by the energy metabolism of fungi, such as pentoses, hydrocarbons, aromatic compounds and others. Lignin is an aromatic hydrocarbon that is only broken down by white rot fungi along with other polysaccharides such as cellulose and hemicellulose. Reducing sugars such as xylose and maltose are present in this hemicellulosic fraction found in agro-industrial waste (Graeme *et al.*, 2018). The results showed that VF and BM provide more reducing sugars, thus characterized by having a free aldehyde group (CH=O) at the end of the chain generating reducing power. Capable of reducing Cu^{2+} into Cu^{+} in Fehling's solution and producing a brick red color. However, in the substrates analyzed there are also small amounts of non-reducing sugars that do not have the ability to undergo oxidation due to the aldehyde functional group being linked to another group, preventing the reducing action. In the Brix° analyses with the diluted samples, it was possible to detect a fraction of 1° Bx for FA and BM, showing the initial availability of diluted sugars that are available and easily assimilated to stabilize the fungus in the substrate and boost the initial enzymatic secretion for the subsequent degradation of more complex structures such as cellulose and lignin. In FA, 0.1 °Bx was detected, showing its low sugar level, in line with the results for total sugars, which were also not detected in the FA samples.

The ash of a material refers to the inorganic residues remaining after the complete destruction of the organic matrix (CO_2 , H_2 O and NO_2). This inorganic matter can contain good amounts of potassium, sodium, calcium, magnesium and small amounts of aluminum, iron, copper, manganese and zinc in the form of oxides, sulfates, phosphates, silicates and chlorides. The values show an ash content of over 1.8%, a value standardized by TACO to refer to samples with a high mineral content. This shows that substrates can also help fungal growth with mineral supply (Moreira *et al.*, 2021). The minerals required for the development of filamentous fungi are trace elements in micromolar concentrations of calcium, copper, iron, manganese and zinc, which are necessary for cell growth (Graeme *et al.*, 2018). AF showed a slightly higher percentage of inorganic material when compared to the other substrates, showing that the substrate has a mineral and nitrogen increment function since sugars were not detected.

As for pH, the substrates proved to be suitable because most fungi are acidophilic and grow between pH 4 and 6. And in a recent fruiting study, it was noted that as the percentage of acid increased from 0.1 to 0.5, there was a tendency to increase primordia, fruiting bodies, biological efficiency and yield (Nwoko *et al.*, 2021). Vilas *et al.* (2020) tested isolates of *Pleurotus* spp. inoculated in different pH ranges and treatments at pH 5 and 6 were superior compared to the other levels studied such as 7, 8 and 9. This demonstrates the suitability of the substrates selected in this study in terms of pH for the species in question.

The moisture content of the isolated substrates on a dry basis was analyzed and VF (22.10%) showed higher moisture content than BM (8.09%) and FA (6.91%). This may be due to the structure of VF, which may have less porous tissue than the other substrates, exhibiting greater water retention capacity, an important characteristic for supporting the development of fungal cells.

Costa *et al.* (2023) showed that the best yields in mushroom production occur with substrates with a higher moisture retention capacity. Thus, the use of VF can contribute to the stabilization of Solid State Fermentation (SSF) by the fungus. Wang *et al.* (2019) mentions the importance of high water content in FES and this parameter is one of the key points in enzymatic activities that play a fundamental role in the effective oxidation of lignocellulosic substrates and consequently for mycelial proliferation.

5.8 C/N ratio of the selected formulations: The carbon content between the different treatments varied from 20.34 to 38.61%. The nitrogen content ranged from 1.19 to 2.96% and the C/N ratio from 13:1 to 26:1. The treatment with the highest carbon content was VF

(38.61%) and the lowest was VF+GA (20.34%), BM+VF obtained 38.17%, BM+VF+GA 29.1%, BM+FA 38.32%, VF+FA 38.38% and V+:FA+GA 33.71% carbon. For the nitrogen content, BM+FA showed the highest value of 2.96% and the lowest content of this element was seen in VF+GA with 1.19%, VF generated 1.45%, BM+VF 2.27%, BM+VF+GA 2.17% VF+FA 2.36% and VF+FA+GA 2.22% of nitrogen in its formulation. The C/N ratio of VF was the highest at 26:1, of VF+GA and BM+VF at 17:1, BM+VF+GA and BM+FA at 13:1, VF+FA at 15:1 and VF+FA+GA at 15:1. These results show that BM can be used to increase the crop's nitrogen levels and consequently reduce the C/N ratio (Table 08).

Table 08. Carbon and nitrogen content and carbon/nitrogen ratio of formulations.

Treatments	C (%)	N(%)	C:N
T2 (VF)	38,61	1,45	26:1
T5 (VF+GA)	20,34	1,19	17:1
T7 (BM+VF)	38,17	2,27	17:1
T10 (BM+VF+GA)	29,1	2,17	13:1
T11 (BM+FA)	38,32	2,96	13:1
T13 (VF+FA)	36,38	2,36	15:1
T14 (VF+FA+GA)	33,71	2,22	15:1

C(%): Carbon; N(%): Nitrogen; C/N: Carbon to Nitrogen ratio. Source> Own authorship, 2024.

The C:N ratio of the substrate is an important factor because it relates its ability to provide carbon and nitrogen to the fungus based on the physiological needs of the species. The ideal range for growing *P. ostreatus* var. *florida* is 7:1 to 40:1 (Hoa *et al.*, 2015 cited by Elbagory et al., 2022). However, matching the C/N ratio alone is not enough because the metabolic behavior of fungi in substrates depends on intricate biochemical and nutritional variables (Feng *et al.*, 2023; Elkhateeb *et al.*, 2022).

5.9 Effect of substrates/treatments on growth time for the production of *Pleurotus ostreatus* var. *florida spawn*: 12 different formulations based on malt bagasse, green bean pods without the grain and algaroba leaves were tested for their potential use in the production of seed inoculum or *spawn*. There was no significant difference between the treatments in terms of colonization time in the different treatments. In all substrates, the mushroom completed colonization between 20 and 25 days (Table 09 and Figures 20 and 21). However, among the substrates tested, the formulations that provided complete colonization in the shortest time were: T6 [Algaroba (100%) + Agricultural Gypsum (2%)], T7 [Malt Bagasse (50%) + Bean Pods (50%)], T8 [Malt Bagasse (50%) + Algaroba (50%)], T10 [Malt Bagasse (50%) + Bean Pods (50%) + Agricultural Gypsum (2%)] and T11 [Malt Bagasse (50%) + Algaroba (50%) + Agricultural Gypsum (2%)]. All these treatments achieved complete colonization in 23 days. When compared to the control group, which showed complete colonization in 20 days and presented high mycelial density and compaction of the grains. The treatments with a 1:1 ratio of malt bagasse to other waste (T7 and T10) followed by treatment T2 (100% bean pods) and T5 (bean pods supplemented with 2% Gypsum) also provided satisfactory development between 24 and 25 days.

The humidity of the different treatments varied from 56 to 78%, as did the vigor and density of the mycelium throughout the colonization period (Table 09 and Figure 18).

Table 09: Analysis of mycelial growth for *spawn* production of *Pleurotus ostreatus* var. *florida*.

Treatments	Total mycelial development time (days)	mycelial development (cm)*	Mycelial growth (cm/day)	U%*	Characteristics of the primary mycelium
TC (control)	20,00 d	11,30 a	0,75 a	56	High mycelial density with a white color and strong compaction of the grains by the mycelium
T1 (BM)	24,00 b	9,00 b	0,60 b	73	High initial density, but as colonization of the substrate progressed, the primary mycelium lost its vigour and acquired a cottony, yellowish appearance
T2 (VF)	24,00 b	7,80 c	0.52 cb	78	Dense, white mycelium,

					characterizing high mycelial vigour
T4 (BM+GA)	24,00 b	9,00 b	0,60 b	68	High mycelial density in new colonies and decreased density of primary mycelium
T5 (VF+GA)	24,00 b	7,60 c	0,50 c	76	Vigorous, compact, white mycelium throughout the colony
T6 (FA+GA)	23,00 c	8.90 cb	0.59 cb	73	Mycelium with medium density
T7 (BM+VF)	23,00 c	10.50 ab	0,70 a	71	High mycelial density and compaction of the substrate
T8 (BM+FA)	23,00 c	9,70 b	0.64 ab	73	Mycelial density and compaction of substrates
T10 (BM+VF+GA)	23,00 c	9,20 b	0.61 ab	70	High mycelial density and compaction of the substrate
T11 (BM+VF+GA)	23,00 c	8.80 cd	0.58 cb	73	Medium density and compaction of the substrate
T13 (VF+FA)	25,00 a	11,00 a	0,73 a	75	High density and compaction of the substrate
T14 (VF+FA)	24,00 b	11,50 a	0,76 a	73	High density and compaction of the substrate
CV%[§]	1,93	2,29	3,46		

U* = Humidity. T1: Malt Bagasse (100%); T2: Green Bean Pods (100%); T4: Malt Bagasse (100%) + Agricultural Gypsum (2%); T5: Green Bean Pods (100%) + Agricultural Gypsum (2%); T6: Algaroba Leaves (100%) + Agricultural Gypsum (2%); T7: Malt Bagasse (50%) + Green Bean Pods (50%); T8: Malt Bagasse (50%) + Algaroba Leaves (50%); T10: Malt Bagasse (50%) + Green Bean Pods (50%) + Agricultural Gypsum (2%); T11: Malt Bagasse (50%) + Algaroba Leaves (50%) + Agricultural Gypsum (2%); T13: Green Bean Pods (50%) + Algaroba Leaves (50%) and T14: Green Bean Pods (50%) + Algaroba Leaves (50%) + Agricultural Gypsum (2%). Averages followed by the same letter do not differ by Duncan's multiple range test ($P<0.05$). Data not transformed.* Humidity of treatments. [*] Average growth up to the 15th day. [§] Coefficient of variation. Source: Own authorship, 2024.

Lignocellulosic substrates that are rich in nitrogen such as malt bagasse (T1) and green bean pods (T2) tend to provide nutritional support for the vegetative growth of *P. ostreatus* var. *florida.* In previous experiments, it was found that malt bagasse alone with agricultural gypsum is ideal for myceliation, however, excess nitrogen could inhibit fruiting in the reproductive phase (Albuquerque, Sousa; 2021), making it necessary to make formulations that add an increase in the Carbon/Nitrogen ratio to malt bagasse in order to improve performance. For *spawn* production, malt bagasse alone (T1) showed unsatisfactory behavior, because as colonization progressed, the growth of the primary mycelium from the top of the polypropylene bag to the end lost vigor, failing to completely sustain a healthy mycelium in the substrate (Figure 19).

Figure 19: Macroscopic aspect of the colonization of *Pleurotus ostreatus* var. *florida* in the different treatments at 15 days after inoculation .

TC: Corn (control); Malt Bagasse (100%); T2: Green Bean Pods (100%); T4: Malt Bagasse (100%) + Agricultural Gypsum (2%); T5: Green Bean Pods (100%) + Agricultural Gypsum (2%); T6: Algaroba Leaves (100%) + Agricultural Gypsum (2%); T7: Malt Bagasse (50%) + Green Bean Pods (50%); T8: Malt Bagasse (50%) + Algaroba Leaves (50%); T10: Malt Bagasse (50%) + Green Bean Pods (50%) + Agricultural Gypsum (2%); T11: Malt Bagasse (50%) + Algaroba Leaves (50%) + Agricultural Gypsum (2%); T13: Green Bean Pods (50%) + Algaroba Leaves (50%) and T14: Green Bean Pods (50%) + Algaroba Leaves (50%) + Agricultural Gypsum (2%). Source: Own authorship, 2024.

On the other hand, the bean pod, when used alone (T2), provided support and ensured a vigorous mycelium until the final colonization time, and the growth/day did not differ significantly from the other treatments. This residue can be considered a source of calcium, phosphorus and nitrogen for the mycelial development of *P. ostreatus* var. *florida* (Pereira *et al.,* 2020).

Thus, formulations containing 50% green bean pods and malt bagasse had a satisfactory mycelial run (T7 and T10) (Table 08). They also provided high compaction of the waste with vigorous mycelium (Figure 20). It is possible that malt bagasse, which has high levels of nitrogen, together with the calcium and phosphorus enrichment of the bean pods, provided the nutritional structure necessary for the growth of the primary mycelium for the production of *P. ostreatus* var. *florida spawn.* The bean pod when formulated with 1:1 algaroba (T13 and T14) also provided support for colonization, demonstrating rapid colonization progress when compared to corn grains (Control Group). Algaroba and malt bagasse (T11) also generated healthy mycelium.

Over the 15-day period, treatments T7, T13 and T14 showed similar growth to the control group (Figure 21). The other treatments showed slower growth, but also proved viable for *spawn* production. Mushroom *spawn* is produced using corn, wheat, sorghum, rice and other grains (Singh *et al.,* 2016).

To be an ideal mushroom spawn for practical use, it should have the following attributes: easy and quick preparation, low space requirements, low cost, rapid mycelium growth and a high yield of fruiting bodies. Ideally, it could also be used after long periods of storage with little or no loss of productivity. This study showed that spawn prepared with the substrates met these requirements, especially if it maintained its fruiting later on. During mushroom cultivation, *spawn* that can be easily and evenly distributed on the substrates is generally considered desirable, because a larger *spawn* area or volume can accelerate the rate at which the mycelium covers the substrate, thus reducing the likelihood of contamination by pathogens such as mold (Zhang et al., 2019).

For rapid colonization, an average of 5% seed is inoculated from the total weight of the substrate. Depending on the grain used, this initial phase can make the production process more expensive. The substrates used in this study are viable for the production of *P. ostreatus* var. *florida spawn* and add value to substrates/waste that would otherwise be discarded.

Figure 20: Evaluation of mycelial growth (cm) in bags 15 days after inoculation of *Pleurotus ostreatus* var. *florida.*

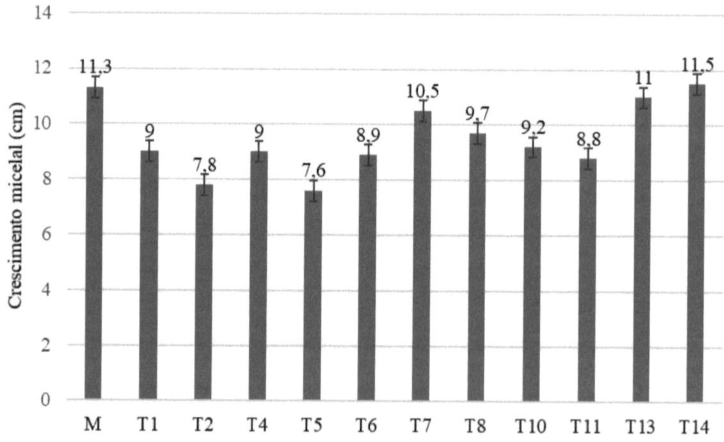

M: maize (control: Malt Bagasse (100%); T2: Green Bean Pods (100%); T4: Malt Bagasse (100%) + Agricultural Gypsum (2%); T5: Green Bean Pods (100%) + Agricultural Gypsum (2%); T6: Algaroba Leaves (100%) + Agricultural Gypsum (2%); T7: Malt Bagasse (50%) + Green Bean Pods (50%); T8: Malt Bagasse (50%) + Algaroba Leaves (50%); T10: Malt Bagasse (50%) + Green Bean Pods (50%) + Agricultural Gypsum (2%); T11: Malt Bagasse (50%) + Algaroba Leaves (50%) + Agricultural Gypsum (2%); T13: Green Bean Pods (50%) + Algaroba Leaves (50%) and T14: Green Bean Pods (50%) + Algaroba Leaves (50%) + Agricultural Gypsum (2%). Source: Own authorship, 2024.

Based on this analysis, 7 treatments were selected (T2, T5, T7, T10, T11, T13 and T14) for the following stages of *Pleurotus ostreatus* var. *florida* production, physicochemical analysis of the basidiomes and N, C and C/N.

5.10. Biological and productive aspects of *Pleurotus ostreatus* var. *florida*: The following treatments were used for productivity tests: T2 (VF), T5 (VF+GA), T7 (BM+VF), T10 (BM+VF+GA), T11 (BM+FA), T13 (VF+FA), T14 (VF+FA+GA). Figure 21 shows the macroscopic appearance of the characteristic basidiomata of *P. ostreatus* var. *florida*, in the 7 previously selected treatments. During production, biological parameters (colonization time, total cycle, stipe size and width, pileus size and humidity) and productive parameters (Yield, Biological Efficiency, Daily Productivity and Compost Consumption) were calculated.

Figure 21. Fruiting of *Pleurotus ostreatus* var. *florida* in the different treatments.

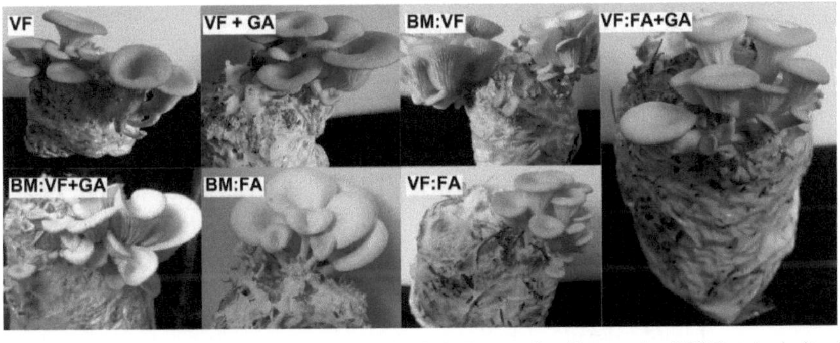

T2 (VF): Bean pods without grains (100%); T5 (VF+GA): Bean pods without grains (100%) + Agricultural Gypsum (2%); T7 (BM+VF): Malt Bagasse (50%) + Bean pods without grains (50%); T10 (BM+VF+GA): Malt Bagasse (50%) + Bean pods without grains (50%) + Agricultural Gypsum (2%); T11 (BM+FA): Malt bagasse (50%) + Algaroba leaves (50%); T13 (VF+FA): Bean pods without grains (50%) + Algaroba leaves (50%); T14 (VF+FA+GA): Bean pods without grains (50%) + Algaroba leaves (50%) + Agricultural gypsum (2%). Source: Own authorship (2024).

Vegetative development (primary mycelium) was satisfactory among all the different formulations and did not differ significantly between treatments. Colonization time (days) and total cycle in all treatments did not exceed 16 and 23 days, respectively. The size/height of the stipes varied from 1.13 to 2.66 cm, with VF (2.16 cm); VF+GA (2.0 cm); BM+VF (2.16 cm); BM+VF+GA (2.66 cm); BM+FA (1.16 cm); VF+FA (1.13 cm) and VF+FA+GA (1.83 cm) and the width varied between 0.23 and 0.70 cm, with VF (0.70 cm); VF + GA (0.6 cm); BM+VF (0.5 cm); BM+VF+GA (0.23 cm); BM:FA (0.55 cm); VF+FA (0.50) and VF+FA+GA (0.26 cm). The size of the pileus ranged from 15.89 to 5.14 cm², with the largest pileus diameter occurring in VF (15.89 cm^2); VF + GA (7.85 cm); BM+VF (5.14 cm); BM+VF+GA (7.66 cm); BM+FA (4.52 cm); VF+FA (4.74 cm) and VF+FA+GA (12.73 cm). The moisture retention capacity of the formulated compounds was assessed and VF showed the highest capacity at 75% moisture, this result being in line with the higher moisture content on a dry basis of the pod, showing that the material really does retain water easily in its structure. VF+GA showed 68% moisture; BM+VF 46%; BM+VF+GA 57%; BM+FA 61%; VF+FA 73% and VF+FA+GA obtained 69% (Table 10).

Table 10. Biological parameters of *Pleurotus ostreatus* var. *florida* found in the first production flow.

Treatments	**Colonization time (days)**	**Total cycle (days)**	**Stipe size (cm)**	**Stipe width (cm)**	**Pileus size (cm)^2**	**U (%)**[1]
T2 (VF)	15	22	2,16±0,16	0,70±0,15	15,89±0,49	75,0
T5 (VF+GA)	15	22	2,00±0,00	0,60±0,05	7,85±0,16	68,0
T7 (BM+VF)	15	22	2,16±0,16	0,50±0,05	5,14±0,23	46,0
T10 (BM+VF+GA)	15	22	2,66±0,33	0,23±0,03	7,66±0,03	57,0
T11 (BM+FA)	15	22	1,16±0,33	0,55±0,16	4,52±0,30	61,0
T13 (VF+FA)	15	22	1,13±0,13	0,50±0,00	4,74±0,23	73,0
T14 (VF+FA+GA)	16	23	1,83±0,16	0,26±0,03	12,73±0,06	69,0
X	15,14	22,14	1,50	0,47	8,36	
(±)EP			0,65	2,05	1,89	
CV(%)			1,10	2,89	2,77	

T: Treatments/Formulations; T2 (VF): Bean pods without grains (100%); T5 (VF+GA): Bean pods without grains (100%) + Agricultural Gypsum (2%); T7 (BM+VF): Malt Bagasse (50%) + Bean pods without grains (50%); T10 (BM+VF+GA): Malt Bagasse (50%) + Grainless Bean Pods (50%) + Agricultural Gypsum (2%); T11 (BM+FA): Malt Bagasse (50%) + Algaroba Leaves (50%); T13 (VF+FA): Grainless Bean Pods (50%) + Algaroba Leaves (50%); T14 (VF+FA+GA): Grainless Bean Pods (50%) + Algaroba Leaves (50%) + Agricultural Gypsum (2%). (±)SE: Standard Error. CV (%): Coefficient of variation. X: Average. Source: Own authorship, 2024.

The influence of the substrates on the number of days from inoculation to complete colonization in the formulations tested showed no significant differences, with an average of 15.14. Rapid colonization also shows productive potential as it prevents contamination of the substrate by other microorganisms such as bacteria, yeasts and fungi that would cause the material to rot and inhibit colonization of the selected species. In the studies by Tavarwisa *et al.* (2021), baobab fruit husks, sawdust and corn cobs were tested and their results showed an average colonization of 25.8, 27.2 and 29.4 days, respectively. Wheat straw was the substrate with the best vegetative colonization rate, which ended in an average of 23 days and a complete total cycle with an average of 31.41 days. The total cycle is the number of days from inoculation to the first harvest and with the compounds tested in this study there was no significant variation, with an average of 22.14 days for mushroom production. In Muswati *et al.* (2021) cotton waste took an average of 18.20 days to complete colonization and a total cycle of 26 days until the first harvest; and compost with cotton waste (50%) and wheat straw (50%) took 21 days to colonize and a total cycle of 31.8 days. A comparison of these results shows that the formulations based on VF, BM and FA are suitable as a growing substrate for white Shimeji and enable production in a short space of time of approximately 22 days.

Also in Muswati *et al.* (2021), the largest hat diameters were obtained with cotton waste (5.5 cm) and with a mixture of baobab fruit peel (50%) and cotton peel (50%) (4.8 cm). In this study, the largest pod diameters were obtained with VF (15.89 cm) and VF+FA (12.73 cm). These results show that algaroba pods and leaves are good nutritional sources for the differentiation of mycelium into mushrooms with high biomass.

In terms of production parameters, a high yield is the main objective in mushroom cultivation as it relates the wet mass of the fresh ba sidiomas to the dry mass of the substrate. In VF this parameter was higher than the other treatments with 154.76%, VF+GA yielded 87.39%, BM+VF 71.05%, BM+VF+ GA 84.44%, BM+VF+GA 84.44%, BM+FA 71.26%, VF+FA 11.01% VF+FA+GA 53.27%.

In terms of Biological Efficiency (%) based on dried mushrooms, in this parameter VF generated 38%, VF+GA yielded 27.73%, BM+VF 37.71%, BM+VF+GA 35.55%, BM+FA 27.58%, VF+FA 29.66% VF+FA+GA 16.38%. In terms of daily productivity (Pr) expressed in grams/day, VF generated 1.45, VF+GA 1.5, BM+VF 1.95, BM+VF+GA 1.45, BM+FA 1.09, VF+FA 1.59, VF+FA+GA 0.86. Finally, in Compost Consumption (CC%) VF showed 73.68, VF+GA 103.47, BM+VF 142.5; BM+VF+GA 80.35, BM+FA 100, VF+FA 122.9, VF+FA+GA 101.66 (Table 11).

Table 11. Production parameters of *Pleurotus ostreatus* var. *florida* shown in the first production flow.

Treatments	U (%)[1]	R (%)[2]	EB (%)[3]	Pr (g/day)[4]
T2 (VF)	75,0	154,76	38,09	1,45
T5 (VF+GA)	68,0	87,39	27,73	1,50
T7 (BM+VF)	46,0	71,05	37,71	1,95
T10 (BM+VF+GA)	57,0	84,44	35,55	1,45
T11 (BM+FA)	61,0	71,26	27,58	1,09
T13 (VF+FA)	73,0	111,01	29,66	1,59
T14 (VF+FA+GA)	69,0	53,27	16,39	0,86
CV*(%)	-	17,9	8,76	3,15

[1]U: Basidium moisture. [2]R: Yield. [3]EB: Biological efficiency. [4]Pr: Productivity. [5]CC: Compost consumption. CV: Coefficient of Variation. Source: Own authorship, 2024.

Mastro *et al.* (2023) carried out a study with wheat straw and found that the biological efficiency varied between 26.28% and 15.49%, decreasing as the storage time of the straw increased. So the results obtained with young wheat straw, the substrate used as a control in mushroom cultivation, are similar to the results obtained with the substrates used in this work, demonstrating that the formulations perform similarly to or better than fresh wheat straw in terms of biological efficiency.

The (CC%) shows the consumption of matter by the fungus, i.e. the degree of colonization of the substrate by the primary mycelium. When the (CC) is higher and associated with high EB, it may mean that there is a possibility that the vegetative mycelium is still active and there is still nutritious organic matter to generate the next flows of basidiocarp production. Thus, the VF, BM+VF and BM+VF+GA formulations with CC values (73.68; 142.50 and 80.35%, respectively), associated with the highest EB values (38.09; 37.71 and 35.55%, respectively) demonstrate the viability of a second production flow after the first harvest due to the availability of substrate that has not yet decomposed and high EB (%) as illustrated in Table 06.

The superior yield results of VF (154.76%), VF + GA (87.39) and VF:FA (111.01%) when compared to those presented by Zakil *et al.* (2019) in which the composite of pressed palm fiber (25%) + sugarcane bagasse (25%) + rubber tree sawdust (50%) generated 123.60 yields in the first flow, 72 in the second flow and 64.62g/kg in the third flow and, in the present study, shows the ability of treatments formulated with VF to produce more than one production flow.

Oliveira *et al.* (2022) also produced *P. eryngii* in axenic blocks, using malt residue (100%) and eucalyptus sawdust supplemented with 10, 15, 20 and 25% malt bagasse (BM) and showed that BM alone did not induce fruiting. However, with supplementation of up to 25%, higher yields were generated in the treatments. In another study by Oliveira *et al.* (2023), malt bagasse (MB) was applied as a nitrogen source in the cultivation of *L. edodes* (Shiitake) and showed that the addition of 35.74% MB to eucalyptus sawdust compost increased the yield to 218g/kg when compared to the addition of only 11% MB, which generated a yield of 46.27 g/kg.

The formulations that generated the largest pileus sizes were those composed of VF (15.83); VF+GA (7.85); VF:BM (7.66) and VF:FA + GA (12.73) (Table 05). And they match the treatments with the best Yields (R%) and Biological Efficiency (EB %) (Table 10), demonstrating the suitability of bean pods without the grain (VF) as a base substrate for the production of *P. ostreatus* var. *florida* due to its ability to guarantee the conversion of the vegetative mycelium into basidiocarp in isolation, i.e. without supplementation.

VF's performance in production can be explained by the presence of sugars that are assimilated by the fungus. Glucose supplementation promotes growth and rapid establishment of the fungus on solid raw material and provides additional easily metabolizable carbon sources. The decrease in reducing sugar values in the substrates after cultivation is associated with their use as an energy source in mushroom production (Zervakis *et al.*, 2022). Malt bagasse and algaroba leaves show greater productive capacity when mixed with VF. Malt bagasse also provides a good amount of assimilable sugars, however the high amount of N in malt bagasse can only favor more fibrous vegetative growth and inhibit fruiting (Bellettini et al., 2019). This makes this substrate more viable as a nitrogen supplement for production when mixed with other substrates. It is possible that algaroba leaves (AF) contain very small amounts of sugars, due to the difficulty in detecting them using the method used in this work, demonstrating that this substrate is viable, but only when mixed with VF or BM.

The highest Yield (154.76%) and Biological Efficiency (38.09%) values were observed in VF alone and in other formulations with VF. EB (%) and R (%) show the degree of suitability of the species in the substrate for generating fungal biomass. In the literature, wheat and corn straw are used as control samples in the cultivation of *P. ostreatus* with yields close to 172% (Akcay *et al.*, 2023). In the present study, yields close to this were found in formulations with VF (154.76%) and VF:FA (111.01%). There are no reports of the use of algaroba leaves in mushroom cultivation in the literature, however, the results of this study show that algaroba leaf biomass (AF) can provide a source of nitrogen and minerals necessary for the development and mycelial differentiation of *P. ostreatus* into basidiomata. Considering that when coupled with BM, a substrate that does not generate basidiomata on its own (Ritota *et al.*, 2019), algaroba leaves also promoted satisfactory fruiting and productivity in treatments formulated with BM.

Thus, in the present study, the BM treatment added productivity when formulated with other substrates and its results are in line with the report by Ritota et al. (2019) in which the use of malt bagasse formulated with other different nitrogen sources such as wheat, rice, corn bran and okara, for the cultivation of *P. ostreatus*, generated greater nutritional value in terms of proteins, carbohydrates and ash in the mushrooms. However, when only malt bagasse not formulated with other substrates was used, low basidiom formation was observed. This may be due to the lack of a specific nutrient in malt cake (MB) necessary for mycelial differentiation into basidiomata.

5.11 Nutritional aspects of the mushroom: The nutritional composition of the mushrooms grown in the different formulations was analyzed and the protein content ranged from 34.35 to 29.15%. VF generated 29.42% protein, VF+GA 29.15%, BM+VF 34.35%, BM+VF + GA 33.5%, BM+FA 34.35%, VF+FA 33.15 and VF+FA+GA 29.15%. The ash content was between 4.92 and 3.76%, with VF at 3.76%, VF + GA 3.76%, BM+VF 4.38%, BM+VF + GA 4%, BM+FA 4.82%, VF+FA 4.92% and VF+FA+GA, 4.5%. The fat content varied between 3.47 and 2.82%, VF obtained 3%, VF+GA 3.47%, BM+VF 2.93%, BM+VF+GA 2.93%, BM+FA 3.06%, VF+FA 3.02% and VF+FA+GA 2,82% fat and the carbohydrate variation was between 50.71 and 59.12% and VF generated 55.12%, VF+GA 59.12%, BM+VF 52.87, BM+VF+GA 53.63, BM+FA 50.71, VF+FA 54.51 and VF+FA+GA 58.13% carbohydrate. And the caloric value at VF was 365.1; VF+GA 380.07; BM:VF 386;

BM:VF+GA 375.8; BM:FA 365.8; VF:FA 369; VF:FA+GA 374 Kcal (Table 12). From the physicochemical analysis of the mushrooms produced, it was observed that the production from treatments mixed with BM generated higher amounts of protein as in BM+VF (34.35%); BM+FA (34.35%), both formulations had higher levels of organic nitrogen with values of 22.75 and 29.63%, respectively.

Table 12. Physicochemical composition of the fruiting bodies of *P. ostreatus* var. *florida* and comparison with Ordinance No. 27 (ANVISA, 2012).

Treatments	Proteins*	Ash*	Fat*	Carbohydrates*	Calories (Kcal)*
T2 (VF)	29,42±0,10	3,76±0,08	3,00±0,26	55,12±0,27	365,15±0,11
T5 (VF+GA)	29,15±0,18	3,76±0,26	3,47±0,10	59,12±0,31	380,07±0,21
T7 (BM+VF)	34,35±0,20	4,38±0,05	2,93±0,09	52,87±0,43	386±0,19
T10 (BM+VF+GA)	33,55±0,67	4,00±0,11	2,93±0,09	53,63±0 ,75	374,89±0,40
T11 (BM+FA)	34,35±0,20	4,82±0,02	3,06±0,22	50,71±0,23	365,8±0,16
T13 (VF+FA)	33,15±0,03	4,92±0,02	3,02±0,24	54,51±0,23	369±0,13
T14 (VF+FA+GA)	29,15±0,13	4,50±0,05	2,82±0,09	58,13±0,39	374±0,16
CV(%)	9,42%	10,76%	20,35%	5,12%	2,26%
Attribute by Ordinance RDC No. 54 of November 12, 2012	High protein content in all formulations in 100g dry basis.		Low in fat, with the exception of the T5 formulation. Maximum 3%		

* Dry basis. CV (%): Coefficient of Variation. Source: Own authorship, 2024.

In this way, it is possible that the N levels in the substrates have increased the protein content of the product and productivity due to the stimulus in the production of enzymes that degrade the substrate. Nitrogen content is one of the main parameters affecting the production of lignocellulolytic enzymes as it is related to the synthesis of proteins, nucleic acid components (purines and pyrimidines), chitin and other polysaccharides that make up the cell wall of many fungi (Melanouri *et al.*, 2022). An increase in nitrogen is observed in mushrooms when related to the amount of the element in the initial substrate and the increase in N in the colonized substrate, showing that the mycelium of *P. ostreatus* can fix nitrogen and use it in the development of the basidiocarp and/or the possible presence of nitrogen-fixing bacteria in symbiosis with the *Pleurotus* genus (Liu *et al.*, 2022; Sun *et al.*, 2021).

In the literature, it is common to find great variation in the macronutrient composition of edible mushrooms due to the use of different substrates in production (Rodrigues *et al.* 2022). The centesimal composition of *P. ostreatus* grown on tucumã, açaí palm, Pará nut shells and pine sawdust was evaluated by Aguiar *et al.* (2021) and found an average of 17.61% protein, 3.03% lipids and 62.03% carbohydrates - with the highest protein content found in the mushroom grown on açaí and tucumã substrate (20.69%). The protein values in the mushrooms analyzed in this work are similar to the protein content of the Korean varieties (28.57%) analyzed by Lee *et al.* (2018) and to the FAO report that shows average protein of the *Pleurotus* species of 28.57% described by Raman *et al.* (2020).

In this study's trials, the treatment with a C/N ratio of 26:1 (VF) produced mushrooms with the largest pileus size (15.89 cm²), the highest EB (%) and the highest R (%). However, in terms of protein, in this study the treatments with a C/N ratio between 13:1 and 17:1 (BM+VF, BM+VF+GA, BM+FA, VF+FA) showed the best performance with a protein content above 30% (Table 12) and did not differ significantly in terms of EB%. The VF+FA treatment stands out when compared to the other treatments as it showed satisfactory results in the two main parameters as it had a good yield (111%) while the protein content (33.15%) was also high and its C/N ratio was 14:1.

Overall, the C/N ratios between 13:1 and 26:1 found in the formulations tested were viable for good productivity and basidiocarp quality. It is also possible to adjust nitrogen levels using substrates that are usually discarded to improve the yield (R%) and protein content of mushrooms for greater product value. In addition, it was also possible to observe the influence of nitrogen on the protein increase in Sassine *et al* (2021), in which the minimum application of doses of nano-urea, a slow-release nitrogen fertilizer applied in the vegetative-primary phase, improved the nutritional attributes of *P. ostreatus,* increasing the protein content with an increase in essential and non-essential amino acids.

In the mushroom's development phase, a lower C/N ratio in the growing substrate is more favorable. However, excess nitrogen, in addition to affecting the formation of fruiting bodies, can also affect the degradation of lignin in *P. ostreatus*, which can impede the development of the mycelium. Using nitrogen above carbon levels can inhibit mycelial growth and delay basidium formation (Bellettini *et al.*, 2019).

In this study, the formulations with the substrates tested produced mushrooms considered to have a high protein content (minimum 12 g of protein per 100 g) and low fat content (maximum 3 g of total fat per 100 g) in accordance with RDC No. 54/2012, which provides for the Mercosur Technical Regulation on Complementary Nutritional Information (Table 6). Consequently, the mushrooms produced showed an ash content above the TACO standardized value of 1.8% to indicate a food rich in minerals and nutritionally safe (Table 6).

Finally, the use of BM in the composition of the substrate for mushroom cultivation increased the nitrogen content, and together with VF promoted greater moisture retention in the substrate and resulted in a greater amount of protein in the mushrooms, which increases consumer acceptance. Thus, the use of VF alone or in formulations with BM and FA supplemented or not with GA, as well as being a way of reusing waste that is difficult to dispose of, represents an alternative for use in the composition of the substrate in the cultivation of the *P. ostreatus* var. *florida* mushroom (Shimeji-white).

6 CONCLUSIONS

The results showed that:

- *P. ostreatus* var. *florida* showed better mycelial development in the dark followed by photoperiod and in the temperature range of 20°C to 25°C.
- Eight formulations [T2 (VF), T6 (FA+GA), T7 (BM+VF), T8 (BM+FA), T10 (BM+VF+GA), T11 (BM+FA+GA); T13 (VF+FA) and T14 (VF+FA+GA)] showed potential for producing *P. ostreatus* var. *florida spawn* and contained bean pods, algaroba leaves and malt bagasse;
- The individual substrates showed satisfactory characteristics for formulations aimed at producing *P. ostreatus* var. *florida* in terms of pH, sugars, ash and soluble solids, favoring fungal growth and fruiting;
- Seven treatments [(T2 (VF), T5 (VF+GA), T7 (BM+VF), T10 (BM+VF+GA), T11 (BM+FA+GA); T13 (VF+FA) and T14 (VF+FA+GA)] provided the necessary nutritional support for basidiom production and the nutritional quality of the mushrooms.
- Treatment T2 composed of VF showed the best biological parameters such as the largest pileus size and the best production parameters such as the highest yield;
- The basidiomes produced in the 7 treatments [(T2 (VF), T5 (VF+GA), T7 (BM+VF), T10 (BM+VF+GA), T11 (BM+FA+GA); T13 (VF+FA) and T14 (VF+FA+GA)] showed low lipid value, high protein content and satisfactory ash, carbohydrate and total caloric value contents.
- The C/N ratio between 13:1 and 26:1 supported the mycelial development and fruiting of *P.ostreatus* var. *florida* in the 7 treatments pre-selected for production.

7 BIBLIOGRAPHICAL REFERENCES

Albuquerque, V. T. N; Sousa, A. C. B. Fungiculture of the edible mushroom *Pleurotus ostreatus* var. *florida*. **Editora Novas Edições Acadêmicas**. Vol. 01, p. 69, 2021.

Alexopoulos, C J.; Mims, C J.; Blackwell, M. Introductory mycology. 4ª ed. New York: **Journal Wiley**, p. 880, 1996.

An, Q; Li, C. S; Yang, J; Chen, S. Y; Ma, K.-Y; et al. Evaluation of laccase production by two white-rot fungi using solid-state fermentation with different agricultural and forestry residues. **BioResources**. Vol. 16, ed. 3, 2021.

Ansiliero, R. et al. Alternatives for Utilizing Fruit Waste - A Review. **Unoesc Videira Research and Extension Yearbook.** Vol. 05, ed. 24976, 2020.

Antunes, P; Corte, L; Cabrera, L; Simões, A; Constantino, L. Market research with mushroom consumers: knowing the preference and profile of those who consume. **International Congress of Agroindustry (CIAGRO): Science Technology and Innovation: from the field to the table**, 2020.

Arango, M; Castañeda, E. Micosis Humanas Procedimentos Diagnósticos. **Direct examinations,** 1995.

Aschemann-Witzel, J; Gantriis, R; Fraga, P; Perez-Cueto, F. Plant-based food and protein trend from a business perspective: markets, consumers, and the challenges and opportunities in the future. **Critical Reviews in Food Science and Nutrition**. Vol, 61, ed 18, 2020.

Association of Official Analytical Chemists (AOAC). Official methods of analysis. 16th ed. Washington, 1995.

Ayimbila, F; Keawsompong, S; Nutritional quality and biological application of mushroom protein as a novel protein alternative. Current Nutrition Reports. Vol, 12, p 290-307, 2023.

Babu, S. et al. Exploring agricultural waste biomass for energy, food and feed production and pollution mitigation: a review. **Bioresource Technology**. Vol. 360, p. 127566, 2022.

Bach, C.V; Helm, M.B; Bellettini, G.M. Maciel, C.W.I. Haminiuk Edible mushrooms: a potential source of essential amino acids, glucans and minerals. **International Journal of Food Science & Technology**. Vol. 52, ed. 11, p. 2382-2392, 2017.

Bellettini, M. et al. Factors affecting mushroom *Pleurotus* spp. **Saudi Journal of Biological Sciences**. Vol. 26, n. 4, p. 13, 2019.

Bernardi, E; Minotto, E; Nascimento, J. Evaluation of growth and production of *Pleurotus* sp. In sterilized substrates. **Agricultural Microbiology**. Vol. 80, n.3, p. 318-324, 2013.

Bisaria, R; Madan, M; Bisaria, V. S. Biological efficiency and nutritive value of *Pleurotus sajor-caju* cultivated on different agro-wastes. **Biological Wastes**. Vol. 19, p. 239-255, 1987.

Bligh, E.G; Dyer,W.J.A. A rapid method of total lipid extraction and purification. Can. J. Biochem. Physiol, v37, n8, p911-917,1953.

Bonatti, M; Karnopp, P; Soares, H.M.; Furlan, S.A. Evaluation of *Pleurotus ostreatus* and *Pleurotus sajor-caju* nutritional characteristics when cultivated in different lignocellulosic wastes. **Food Chem.**, 88(3): 425-428, 2004.

Brazil. Ministry of Health. Resolution RDC no. 360, of December 23, 2003. Technical Regulation on nutritional labeling of packaged foods. **ANVISA - National Health Surveillance Agency**, Virtual Health Library: Brazil, DF, 2003.

Chang, S.T; Lau, O.W; Cho, K.Y. The cultivation and nutritional value of *Pleurotus sajor-caju*. European Journal Microbiology Biotechnology. Vol. 12, p. 58-62, 1981.

Coelho, J; Ximenes, L. Beans: production and market. **Caderno Setorial - Escritório Técnico de Estudos Econômicos do Nordeste (ETENE)**. vol. 05, ed. 143, 2020.

Colla, I.M et al. Carbon-to-nitrogen ratios on laccase and mushroom production of *Lentinus crinitus*. **International Journal of Environmental Science and Technology**. Vol. 20, p. 3941-3941, 2023.

Costa, A. et al. The use of rice husk in the substrate composition increases *Pleurotus ostreatus* mushroom production and quality. **Scientia Horticulturae**. Vol. 321, 2023.

Dantas, E. Symbiosis between algaroba and rhizobia naturally established in soils of Pernambuco: potential for nitrogen contribution. **Postgraduate Program in Energy and Nuclear Technologies (Thesis)**. Federal University of Pernambuco, Recife, 2022.

Daud, M. Habitat characteristics and utilization of edible wild mushrooms by local communities in the protected forest in Pinrang Regency, Indonesia. **Earth and Environmental Science**. Vol. 886, 2021.

Desisa, B; Muleta, D; Dejene, T; Jida, M; Goshu, A. Martin-Pinto. Substrate optimization for shiitake (*Lentinula edodes* (Berk.) Pegler) mushroom production in Ethiopia. **J.Fungi**. vol. 9. ed. 8, 2023.

Díaz-Godínez, G; Téllez, M. Mushrooms as edible foods. **Fungi in Sustainable Food Production, Fungal Biology.** 2021.

Dilfy, S; Hanawi, A; Jalil, A. Determination of chemical composition of cultivated mushrooms in Iraq with spectrophotometrically and high-performance liquid chromatographic. **Journal of Green Engineering (JGE**). Vol. 10, ed. 09, 2020.

Doroski, A. et al. Food waste originated material as an alternative substrate used for the cultivation of oyster mushroom (*Pleurotus ostreatus*): a review. **Sustainability: New Food** Waste Horizons. Vol. 14, p. 19, 2022.

Elkhateeb, W; Daba, D. Mushrooms as efficient enzymatic machinery. **Journal of Biomedical Research & Environmental Sciences**. Vol. 03, ed. 04, p. 423-428, 2022.

Feng, Y; Xu, H; Sun, Y; Xia, R; Hou, Z; Wang, Y; Pan, S; Li, L; Zhao, C; Ren, H; Xin, G. Effect of light on quality of preharvest and postharvest edible mushrooms and its action mechanism: a review. **Trends in Food Science & Technology**. Vol. 139, 2023.
Fermor, T. R; Wood, D. A. Mushroom compost microbial biomass: a review. **Mushroom Science**, Leamington Spa,. Vol. 13, p. 191-199, 1991.

Fornito, S. Puliga, F. Leonardi, P. Degradative ability of mushrooms cultivated on corn silage digestate. **Molecules - Mushrooms: The Versatile Roles**. Vol. 25, ed. 13, p. 3020, 2020.

Fox, G. P Chemical composition in barley grains and malt quality. Department of Primary **Industries & Fisheries**. 2010. Available at: https://www.researchgate.net/figure/Structure-of-barley_fig1_227235861. Accessed on: Jan 2024.

Furlan, S A; Virmond, L J; Miers, D; Bonatti, M; Gern, R M M; Jonas, R. Mushroom strains able to grow at high temperatures and low pH values. **World Journal of Microbiology and Biotechnology**. Vol. 13, p.689-692, 1997.

Gomes, F. P. Curso de Estatística Experimental. Piracicaba: Nobel, ed.13, 1990, p. 476, 1990.

González, A. et al. Evaluation of functional and nutritional potential of a protein concentrate from Pleurotus ostreatus mushroom. **Food Chemistry**. vol 346, ed 1, 2021.

Grimm, D; Kuenz, A; Rhmann, G. Integration of mushroom production into circular food chains. **Organic Agriculture**. Vol. 11, p. 25-36, 2021.

He, J; Evans, N.M; Liu, H; Shao, S. A review of research on plant-based meat alternatives: driving forces, history, manufacturing, and consumer attitudes. Compr. Rev. **Food Sci**. Food Saf. Vol. 19, p. 2639-2656, 2020.

Hoa, H.T; Wang, C.L; Wang, C.H. The effects of different substrates on the growth, yield, and nutritional composition of two oyster mushrooms (*Pleurotus ostreatus* and *Pleurotus cystidiosus*). **Mycobiology**. Vol. 43, p. 423-434, 2015.

Holtz, M. Utilization of cotton waste from the textile industry for the production of fruiting bodies of *Pleurotus ostreatus* DSM 1833. **Journal of Environmental Sciences**. Vol. 3, p. 37-51, 2009.

Ibrahim, R; Ali, M; Abdel-Salam, F. Nutritional and quality characteristics of some foods fortified with dried mushroom powder as a source of vitamin D. **Inter Journal of Food Science**. vol. 22, 2022.

Imtiaj, A; Jayasinghe, C; Lee, G. W; Kim, H. Y; Shim, M. J; Roh, H. S; Lee, H. S; Hur, H; Lee, M. W; Lee, U. Y; Lee, T. S. Physicochemical requirement for the vegeta-tive growth of

Schizophyllum commune collected from different ecological origins. **Mycobiol.,** v.36, p.34-39, 2008.

Inci, S. et al. Antimicrobial, antioxidant, cytotoxicity and DNA protective properties of the pink oyster mushroom, *Pleurotus djamor* (Agaricomycetes). **Inter Journal of Medicinal Mushroom.** Vol. 25, n. 2, p. 55-66, 2023.

Adolfo Lutz Institute. Analytical standards, chemical and physical methods for food analysis. Ed. 3, São Paulo: **Adolfo Lutz Institute.** Vol. 1, p. 533, 1985
Jedvert, K; Heinze, T. Cellulose modification and shaping - a review. **Journal of Polymer Engineering**. Vol. 37, ed. 09, 2017

Jiang, W; Pei, R; Zhou, S.F. 3D-printed xylanase within biocompatible polymers as excellent catalyst for lignocellulose degradation. **Chem. Eng. J**. Vol. 400, ed. 125920, 2020

Kalili, A. Ouafi, R. Naciri, K. Chemical composition and antioxidant activity of extracts from Moroccan fresh fava beans pods (*Vicia faba* L.). **Baza Agro**. Vol. 73, ed. 1, 2022.

Kuijk, V.S; Sonnenberg, A; Baars, J; Hendriks, W; Cone; J. Fungal treated lignocellulosic biomass as ruminant feed ingredient: a review. Biotechnology advances. Vol. 33 , p. 191 - 202, 2015

Kumla, J; Sunwannarch, N; Sujarit, K; Penkrue, W. Cultivation of mushrooms and their lignocellulolytic enzyme production through the utilization of agro-industrial waste. Molecules - **Mushrooms: The Versatile Roles**. Vol. 25, ed. 12, p. 2811, 2020.

Food Technology Laboratory. **Manual of Food Analysis Practices.** João Pessoa: 2020, 57p.

Lima, L; Silva, J; Cavalcante, L; Silva, E. Evaluation of the adsorptive power of bean pods (*Phaseolus vulgaris* L.) in bodies of water contaminated with gasoline, using the adsorption technique. **CONIMAS: 1st International Congress on Seminar Diversity**. Vol. 03, 2019.

Liu, Q. et al. Dynamic succession of microbial compost communities and functions during Pleurotus ostreatus mushroom cropping on a short composting substrate. **Frontiers in Microbiology**, Vol. 13, 2022.

Loh, H; Seah, Y; Looi, I. The Covid-19 pandemic and a diet change. **Progress in microbes and molecular biology**. Vol. 4, ed. 1, 2021.

Lopes, R. S; Portela, A. P. A. S; Svedese, V. M.; Albuquerque, A. C; Lima, E. A. L. A. Morphological aspects of *Paecilomyces farinosus* (HOLM EX S. F. GRAY) BROWN and SMITH on infection in *Coptotermes gestroi* (WASMANN) (ISOPTERA: RHINOTERMITIDAE). **Biological.** Vol. 70, p. 29-33, 2008.

Majesty, D; Ijeoma, E; Winner, K; Prince, O. Nutritional, anti-nutritional and biochemical studies on the oyster mushroom, *Pleurotus ostreatus*. EC-Nutrition. Vol. 16, 2018.

Marçal, S; Sousa, A; Taofiq, O. Antunes, F. Morais, A. Freitas, A. Barros, L. Ferreira, I. Pintado, M. Impact of postharvest preservation methods on nutritional value and bioactive properties of mushrooms. **Trends in Food Science & Technology**. vol, 110, p. 418-431, 2021.

Martin, C; Parlasca, Q; Martin. Meat consumption and sustainability. **Annu. Rev. Resour. Econ**. 2022, 14, 17-41.

Massardi, M; Massini, M; Silva, D. Chemical characterization of malt bagasse and evaluation of its potential for obtaining value-added products. **The Journal of Engineering and Exact Science**. Vol. 06, ed. 01, 2020.

Mastro, F. Traversa, A. Matarrese, F. Cocozza, C. Brunetti, G. Influence of growing substrate preparation on the biological efficiency of *Pleurotus ostreatus*. **Horticulturae**. Vol. 09, ed. 04, 2023.

McKillop, K; Harnly, J; Pehrsson, P; Fukagawa, N; Finley, J. FoodData Central, USDA's updated approach to food composition data systems. **Curr Dev Nutr**. Vol. 5, ed 5:596, 2021.

Melanouri, E; Dedousi, M; Cultivating *Pleurotus ostreatus* and *Pleurotus eryngii* mushroom strains on agro-industrial residues in solid-state fermentation. Part I: Screening for growth, endoglucanase, laccase and biomass production in the colonization phase. **Carbon Resources Conversion**. Vol. 05, ed. 01, p 61-70, 2022.

Mendes, C; Fleming, R; Cena, C. Mechanical and microstructural characterization of epoxy/sawdust (*Pinus elliottii*) composites. **Sage Journals.** vol. 29, ed. 8, 2021.

Miles, P. G; Chang, S. T. Biologia de las setas: fundamentos básicos y acontecimientos actuales. **World Scientific**, Hong Kong, 1997.

Mohamad Mazlan, M; Talib, RA; Chin, NL; Shukri, R; Taip, FS; Mohd Nor, MZ. et al. Physical and microstructure properties of oyster mushroom-soy protein meat analog via single-screw extrusion. **Foods**. Vol, 09, ed 08, p. 1023, 2020.

Mora, O; Mouël, C; Lattre-Gasquet, M; Donnars, C; Dumas, P; Réchauchère, O; Brunelle, T; Manceron, S; Marajo-Petitzon, E; Moreau, C; Barzman, M; Forslund, A; Marty, P. Exploring the future of land use and food security: a new set of global scenarios. **Plos One**. Vol. 15, ed. 7, 2020.

Moreira, D; Dias, T; Rocha, V; Chaves, A. Determination of ash content in foods and its relationship with health. **Ibero-American Journal of Humanities, Sciences and Education - REASE**. São Paulo, Vol. 7, ed.10, 2021.

Moretto, E; Fett, R; Gonzaga, L.V; Kuskoski, E.M. 2002a. Composição centesimal dos plantas e outros materiais. 2ª ed. UFRGS, 174 p. **Food products.** Chapter 1. *in:* Moretto *et al.* (eds). *Introduction to food science*.

Moura, M; Martins, B; Oliveira, G; Takahashi, J. Alternative protein sources of plant, algal, fungal, and insect origins for dietary diversification in search of nutrition and health. **Food Science and Nutrition.** Vol. 63, ed. 31, 2023.

Muswati, C. et al. The effects of different substrate combinations on growth and yield of oyster mushroom (*Pleurotus ostreatus*). **International Journal of Agronomy**. Vol. 21, p. 10, 2021.

Navarro, M; Gea, F; Gimenez, A; Martinez, A. Raz, D. Levanon, D. Offer, D. Agronomical valuation of a drip irrigation system in a commercial mushroom farm. **Scientia Horticulturae**. Vol. 265, ed. 30, 2020.

Nogueira, A.R; Souza, G. B. Manual de laboratórios: solo, água, nutrição vegetal, nutrição animal e alimentos. São Carlos: **Embrapa Pecuária Sudeste**, 2005. 334p.

Nwoko, M; Achufusi, J; Ehumadu, C; Okere, C. Ukpai, K. Influence of pH on the fructification, some macro-morphological characters and productivity of *Pleurotus ostreatus* (Jacq: Fr) Kumm. sporophores cultivated on HCl-induced substrate. **International Journal of Advanced Academic Research**. Vol. 7, ed. 11, 2021.

Oliveira, G; Cuquel, F; Shiitake cultivation in axenic blocks containing brewery residue. **Brazilian Journal of Agricultural Sciences**. Vol. 18, ed. 1, 2023.

Oliveira, G; Jesus, G; Silva, R; Cuquel, F. King oyster mushroom production in axenic blocks supplemented with brewery residue. **Acta Hortic.** International Society for Horticultural Science. vol. 1355, p. 363-368, 2022.

Oosone S; Kashiwaba A; Yanagihara, N; Yoshikawa, J; Kashiwagi, Y; Maehashi, K. et al. The role of amylolytic and proteolytic enzyme activities of vegetables, fruits, and edible fungi in flavor enhancement during cooking. **Int J Gastron**. Vol. 22, ed. 100264, 2022.

Parvin, R; Farzana, T; Mohajan, S. Rahman H, Rahman SS. Quality improvement of noodles with mushroom fortified and its comparison with local branded noodles. **NFS J.** Vol 26, ed 08, 2020.

Pedra, W. N; Marino, R. H. Axenic cultivation of *Pleurotus* spp. on sawdust of coconut (*Cocos nucifera* Linn.) bark supplemented with and/or rice bran. Arch. **Inst. Biol.** Vol.73, n.2, p.219-225, 2006.

Pellegrino, R; Blasi, F; Angelini, P; Ianni, F. LC/MS Q-TOF metabolomic investigation of amino acids and dipeptides in *Pleurotus ostreatus* grown on different substrates. **J. Agric. Food. Chem**. Vol. 70, ed. 33, p. 103711-10382, 2022.

Qin, F; Johansen, A. Z; Mussatto, S. I. Evaluation of different pretreatment strategies for protein extraction from brewer's spent grains. **Industrial Crops and Products**, Vol. 125, n. June, p. 443-453, 2018.

Ricardino, I; Souza, M. Neto, I. Vantagens e Possibilidades do Reaproveitamento De Resíduos Agroindustriais. **Environment and Food**. Vol. 01, ed.08, 2020.

Ritota, M; Manzi, P. *Pleurotus* spp. cultivation on different agri-food by-products: example of biotechnological application. **Sustainability - MDPI**. Vol. 11, ed. 5049, 2019.

Rodrigues, G; Okura, M. Edible mushrooms in Brazil: a literature review. Research, **Society and Development**, Vol. 11, n. 8, 2022.

Rugolo, M.; Lechner, B.; Mansilla, R.; Gerardo, M.; Rajchenberg, M. Evaluation of *Pleurotus ostreatus* basidiomes production on Pinus sawdust and other agricultural and forestry wastes from patagonia, Argentina. Maderas, **Cienc. Tecnol**, Concepción, Vol. 22, p. 517-526, 2020.

Salami, A. O; Bankole F. A; Olawole, I. O. Effect of different substrates on the growth and protein content of oyster mushroom (*Pleurotus florida*). Int. J. **Biol. Chem. Sci.** Vol. 10, ed. 2, p. 475-485, 2016.

Santos, P; Alexandre, F; Guedes, M; Lucena, F; Cavalcanti, M. Use of the algarobeira (*Prosopis juliflora*) (Sw.) DC) in the semi-arid region: the case of Ribeira do Riacho do Navio. **Forúm Ambiental da Alta Paulista**. Vol. 06, ed. 04, 2020.

Sassine, Y. et al. Nano urea effects on *Pleurotus ostreatus* nutritional value depending on the dose and timing of application. **Sci Rep**. Vol. 11, 2021.

Sirmah, P. Are the pods and leaves of *Prosopis juliflora* growing in Baringo Kenya toxic to livestock chemical analysis perspectives. **Asian Journal of Natural & Applied, Sciences**. Vol. 7, ed. 3, 2018.

Sun, C. et al. Improve spent mushroom substrate decomposition, bacterial community and mature compost quality by adding cellulase during composting. **Journal of Cleaner Production**. Vol. 299, 2021.

Sun, X; Dou, Z; Shurson, G; Hu, B. Fungal bioprocessing of wheat straw with fruit and vegetable discards to produce cattle feeds for enhanced sustainability. **Resources, Conservation and Recycling**. Vol. 199, 2023.

Tagkouli, D; Kaliora, A; Bekiaris, G; Kourotsios, G; Christea, M; Zervakis, G; Kalogeropoulos, N. Free amino acids in three Pleurotus species cultivated on agricultural and agro-industrial by-products. **Molecules**. Vol. 25, ed. 17, 2020.

Tavarwisa, D; Govera, C; Moses, M; Ngezimana, W. Evaluating the Suitability of Baobab Fruit Shells as Substrate for Growing Oyster Mushroom (*Pleurotus ostreatus*). **International Journal of Agronomy**, V 2021, P 7, 2021.

Tavarwisa, M; Govera, C; Mutetwa, M; Ngezimana, W. Evaluating the suitability of baobab fruit shells as substrate for growing oyster mushroom (*Pleurotus ostreatus*). **International Journal of Agronomy.** Vol. 2021, p. 7, 2021.
Tedesco, M.J; Gianello, C; Bissani, C.A; Bohnen, H; Volkweiss, S.J. 1995. **Soil analysis**, UFSC, pg. 19-56.

Teixeira, P. C; Donagemma, G. K; Fontana, A; Teixeira, W. G. Solo: Manual de Métodos de Análise de Solo Embrapa Solos. p. 577, 2017.

Thakur, M.P. Advances in mushroom production: key to food, nutritional and employment security: a review. **Indian Phytopathology**, Vol. 72, 2020.

Tombini, C; Vrgas, P.F; Feltes, M. M.C; Lajus, C.R. Centesimal, Physico-Chemical, Bioactive and Microbiological Characterization of Pilsen-Type Malt Bagasse. **Journal of Social and Environmental Management.** Vol. 18, ed. 1, 2024.

Vilas, P; Jadhav, A; Dhavale, M. Hasabnis, S. Effect of cultural variability on mycellial growth of eleven mushroom isolates of Pleurotus. Journal of Pharmacognosy and Phytochemistry. vol. 9, ed. 6, 2020.

Wang, F; Xu, L; Zhao, L; Ding, Z; Ma, H; Terry, N. Fungal laccase production from lignocellulosic agricultural wastes by solid-state fermentation: a review. **Microorganism**. Vol. 7, ed. 12. p.665, 2019.

Yu, Q; Guo, M; Zhang, B; Wu, H; Zhang, Y; Zhang, L. Analysis of nutritional composition in 23 kinds of edible fungi. **J Food Qual.** Vol. 8821315, 2020

Yuan, X; Jiang, W; Zhang, D; Liu, H; Sun, B. Textural, sensory and volatile compounds analyses in formulations of sausages analogue elaborated with edible mushrooms and soy protein isolate as meat substitute. **Foods.** Vol. 11, ed. 1:52, 2021

Zakil, F; Hassan, M; Sueb, M; Isha, R. Growth and yield of *Pleurotus ostreatus* using sugarcane bagasse as an alternative substrate in Malaysia. **Materials Science and Engineering**. Vol. 736, 2020.

Zakil, F; Sueb, S; Isha, R. Growth and yield performance of *Pleurotus ostreatus* on various agro-industrial wastes in Malaysia. **Proceedings of the 2nd International Conference on Biosciences and Medical Engineering (ICBME2019)**. Vol. 215, 2019.

Zanetti, A. L; Ranal, M. A. Supplementation of sugarcane with guandu in the cultivation of *Pleurotus* sp. *florida*. **Pesquisa Agropecuária Brasileira, Brasília**. Vol. 32, n. 9, p. 959-964, 1997.

Zervakis, G; Martin, C; Xiong, S; Koutotsios, G; Straetkvern, K. Spent substrate from mushroom cultivation: exploitation potential toward various applications and value-added products. **Bioengineered**, Vol. 14, 2022.

Zhang, W; Liu, S; Kuang, Y; Zheng, S. Development of a novel spawn (block spawn) of an edible mushroom, *Pleurotus ostreatus*, in liquid culture and its cultivation evaluation. **Mycobiology**. Vol. 47, ed. 1, 2019.

Zheng, Y; Yichun, X; Xie, Y; Yu, S. Asexual reproduction and vegetative growth of Bionectria ochroleuca in response to temperature and photoperiod. **Ecology and Evolution**. Vol. 11, p. 10515-10525, 2021.

Zhuo, R; Yu, H; Qin, X; Ni, H; Jiang, Z.; Ma, F.; Zhang, X. Heterologous expression and characterization of a xylanase and xylosidase from white rot fungi and their application in synergistic hydrolysis of lignocellulose. **Chemosphere**. Vol. 212, p. 24-33, 2018.

yes

I want morebooks!

Buy your books fast and straightforward online - at one of world's fastest growing online book stores! Environmentally sound due to Print-on-Demand technologies.

Buy your books online at
www.morebooks.shop

Kaufen Sie Ihre Bücher schnell und unkompliziert online – auf einer der am schnellsten wachsenden Buchhandelsplattformen weltweit! Dank Print-On-Demand umwelt- und ressourcenschonend produzi ert.

Bücher schneller online kaufen
www.morebooks.shop

info@omniscriptum.com
www.omniscriptum.com

Printed by Books on Demand GmbH, Norderstedt / Germany